T0326499

Frameworks for Scientific and Technological Research Oriented by Transdisciplinary Co-Production

Frameworks for Scientific and Technological Research Oriented by Transdisciplinary Co-Production

Prof. Lillian Maria Araujo de Rezende Alvares, PhD
Prof. Patricia de Sá Freire, PhD

ANTHEM PRESS

Anthem Press
An imprint of Wimbledon Publishing Company
www.anthempress.com

This edition first published in UK and USA 2023
by ANTHEM PRESS
75–76 Blackfriars Road, London SE1 8HA, UK
or PO Box 9779, London SW19 7ZG, UK
and
244 Madison Ave #116, New York, NY 10016, USA

British Library Cataloguing-in-Publication Data
A catalogue record for this book is available from the British Library.

Library of Congress Cataloging-in-Publication Data
A catalog record for this book has been requested.

ISBN-13: 978-1-83998-6-840 (Hbk)
ISBN-10: 1-83998-6-840 (Hbk)

This title is also available as an e-book.

ACKNOWLEDGEMENTS

We would like to thank our Laboratory of Integration Engineering and Multilevel Governance of Knowledge and Organizational Learning (ENGIN), the Graduate Program in Engineering and Knowledge Management (EGC), the Federal University of Santa Catarina (UFSC/Brazil) and the Coordination of Higher Education Personnel (CAPES), foundation of the Ministry of Education (MEC/Brazil) for supporting the research that gave rise to this book.

CONTENTS

LIST OF FIGURES

LIST OF TABLES

Human inconstancy is to be blamed. Nobody is only one thing, we are all many. And the worst thing is that from one side of us one cannot deduce the other, right?[1]

Luis Fernando Verissimo, 2013

1 *A culpa é da inconstância humana. Ninguém é uma coisa só, nós todos somos muitos. E o pior é que de um lado da gente não se deduz o outro, não é mesmo?*

INTRODUCTION

Transdisciplinarity is a relatively recent term: it emerged in 1970 in France during an event held to discuss the role of multidisciplinarity and interdisciplinarity in university settings, entitled *'L'interdisciplinarité: problèmes d'enseignement et de recherche dans les universités'*.[1,2] During the event, three participants presented and discussed the new concept of transdisciplinarity. Jean Piaget and Andre Lichnerowicz focused on disciplinary relations, while Erich Jantsch addressed the concept from the perspective of social purpose.

The event proceedings (Apostel et al. 1970), with more than 300 pages, were the most important source of reference in the field for a long time, until two other theoretical frameworks were published (Klein 2013): Mode 2 Knowledge Production of Michael Gibbons, Limoges, Nowotny, Schwartzman, Scott and Trow (1994) and Basarab Nicolescu's Manifesto of Transdisciplinarity (1993).

All perspectives, moreover, sought to understand the multidimensionality of reality and the inclusion of social values that dismantle the division into academics (i.e., the experts) and non-academics, fostering new partnerships between university and society. The valorization and increment of these forms of interaction in the pursuit of the unity of knowledge signalled a new form of integration between society and academia, especially for the conduct of scientific research.

The pursuit of the unity of knowledge, in fact, is the epistemological problem highlighted by transdisciplinarity. It dates back to the time of ancient philosophers and has never ceased to be a relevant topic of study. In Morin's words, it flows through time:

> As Pascal said: 'I hold it is as impossible to know the parts without knowing the whole as to know the whole without knowing the particular parts'.

1 Interdisciplinarity: teaching and research issues in universities.

2 Held by the organization for Economic Cooperation and Development (OCDE), in collaboration with the Ministry of Education and the University of Nice, from 7 to 12 September 1970.

> Pascal's statement reminds us of the need for back-and-forth movements that run the risk of forming a vicious circle, but which can constitute a productive circuit as in a shuttle movement that weaves the development of thought [...] complexity is not just the union of complexity and non-complexity (simplification); complexity is at the heart of the relationship between the simple and the complex, because such a relationship is antagonistic and complementary.[3] (Morin 2005, p. 136)[4]

In this context, transdisciplinarity is involved in a series of changes that have become necessary to transcend and integrate disciplinary paradigms. It is, as described by Klein (2009), the shift from fragmentation to relationality, from unity to the integrative process, from homogeneity to heterogeneity, from linearity to non-linearity, from simplicity to complexity, from universality to situated practices, from isolation to collaboration and cooperation.

In order to deepen the above aspects with a view to highlighting the search for scientific knowledge through transdisciplinary research, this work aims to offer a historical and epistemological vision of the scope of transdisciplinarity, including the conceptual frameworks of transdisciplinary research. Finally, with the scenario presented, the authors propose a theoretical and methodological referential framework entitled Knowledge Acquisition Design (KAD), a framework for the development of scientific and technological research of transdisciplinary co-production.

This work is divided into six chapters. Chapter 1 seeks to find the roots of the concern with the unity of knowledge, a founding element of transdisciplinarity, from the Renaissance onwards, but including the contributions of philosophers of classical antiquity. Chapter 2 is explicitly devoted to transdisciplinarity, especially historical and epistemological milestones. Chapter 3 is devoted to the transdisciplinary co-production of scientific and technological knowledge and will explore the meaning

---

3 *Comme disait Pascal: 'Je tiens pour impossible de connaître les parties en tant que parties sans connaître le tout, mais je tiens pour non moins impossible la possibilité de connaître le tout sans connaître singulièrement les parties'. La phrase de Pascal nous renvoie à la nécessité des va-et-vient qui risquent de former un cercle vicieux mais qui peuvent constituer un circuit productif comme dans un mouvement de navette qui tisse le développement de pensée. [...] la complexité, ce n'est pas seulement l'union de la complexité et de la non-complexité (la simplification); la complexité est au coeur de la relation entre le simple et le complexe parce qu'une telle relation est à la fois antagoniste et complémentaire.*

4 The first edition of *Introduction à la pensée complexe* was published in 1990.

per se and evolution of the term 'co-production'. Chapter 4 presents selected conceptual and methodological frameworks for transdisciplinary research conducted from 2005 to 2019. Building on the previous segments, Chapter 5 introduces the KAD framework to conduct transdisciplinary co-production research on knowledge governance and organizational learning. Finally, Chapter 6 summarizes the contents of this book and lists the definition of the main concepts addressed previously.

Chapter 1

THE CONCERN WITH THE UNITY OF KNOWLEDGE IN HISTORY

Reflections on the unity of knowledge were widely expressed at different times in history. They started with the philosophers of antiquity, and then continued in the Middle Ages, during the Enlightenment, in the theoretical foundations of the mereology of systemic thought and also throughout the philosophy of science. Debate on the theme was intensified in the sixteenth century and later with the revolutionary ideas of German, French, English and Italian thinkers about knowledge and the scientific method that will give rise to modern science; for example, Francis Bacon, Galileo Galilei, René Descartes, Blaise Pascal, John Locke, Isaac Newton, Gottfried W. Leibniz, David Hume, and Immanuel Kant. Table 1.1 shows how the ideas of such thinkers intersect throughout time.

In the wake of modern science, however, the issue of disciplinary specialization has emerged, causing a profound change in the search for scientific knowledge. The new order of increasingly specialized disciplines negatively affected the integrated view of science and knowledge, as it kept a distance from the various realities found in the same problem or circumstance and even disregarding the necessary integration between different types of knowledge. Upon realization of this trend, numerous movements and initiatives emerged out of the concern for the breakdown of the unity of knowledge. The Enlightenment philosophers, for example, organized the Encyclopédie or Dictionnaire Raisonné des Sciences, des Arts et des Métiers (Diderot and D'Alembert 1751–1772; *Encyclopedia, or a Systematic Dictionary of the Sciences, Arts, and Crafts*), a collection of 17 volumes, released between 1751 and 1772 in order to gather, organize and disseminate the knowledge available. Despite this grand undertaking, science has been fragmented, and this situation remains until today.

In the 1990s, however, the debate on the importance of pursuing the unity of knowledge was rekindled, enhanced by the concept of transdisciplinarity, introduced in the 1970s. It was argued that the traditional mode of knowledge production (Mode 1) gave way to a new way (Mode 2) that was more suited

Table 1.1 Birth and death years of some thinkers whose legacy has consolidated modern science

Thinker	Century																	
	16th			17th							18th						19th	
Bacon	1561				1626													
Galilei		1564					1642											
Descartes			1596							1650								
Pascal				1623							1662							
Locke						1632						1704						
Newton								1643								1727		
Leibniz									1646					1716				
Hume													1713				1784	
Kant															1724			1804

Source: The authors (2021).

to capturing the complexity of reality through scientific research. In a nutshell, for the mentors of the idea, Gibbons et al. (1994), Mode 1 is mostly applied to academic issues of a specific community, unlike Mode 2, which operates in context, wherever it is; Mode 1 is homogeneous but Mode 2 is heterogeneous; Mode 1 is hierarchical and constant while Mode 2 is heterarchical and transient; Mode 1 is disciplinary while Mode 2 is transdisciplinary.

To acknowledge the importance of exploring all aspects of the same situation, this chapter will take a deeper look into ideas and views about the unity of knowledge – from some of the first thinkers who addressed the issue up to Kurt Gödel, who is considered to be the father of transdisciplinarity.

1.1 The New Science Described by Francis Bacon (1561–1626)

Francis Bacon (1561–1626) devoted much of his intellectual energy to explaining the new science of the seventeenth century. In his seminal work *Advancement of Learning* (Bacon 1605) he argued that deductive logic (or Aristotelian logic) of the quest for truth and knowledge, which had originated in antiquity,[1] was insufficient for the development of science in the modern era. Bacon systematized the inductive method, which was widely used not only by his contemporary Galileo (1564–1642) in an innovative combination of experiments and mathematics, but also by others even before these major figures of science and the scientific method. Lakatos and Marconi (1991) stated that 'induction, as a reasoning technique, has existed since Socrates and Plato'[2] (p. 64).

These two forms of reasoning – deduction and induction – are rational methods for understanding science. Instead of just following thought freely, deductive and inductive methods are based on reflection; they follow a coherent procedure, based on elements of reason. The former, named as syllogism by

1 Aristotelian logic (or the deductive method) for understanding reality and pursuing knowledge was developed in classical antiquity and presented in the Organon, which, *roughly speaking*, concerns the following works by Aristotle: Categories, On Interpretation, Prior Analytics, Posterior Analytics (or both just called Analytics by the philosopher), Topics, and On Sophistical Refutations (the latter being the final section of Topics). The thinker himself did not use the name Organon to refer to the set (which was only grouped under that name in the Middle Ages). Aristotle did not even treat them as parts of a single work. Many researchers also associate the work Rhetoric to the set, as a confirmation of Topics (Stanford Encyclopedia of Philosophy 2017a).

2 Our translation.

Aristotle, departs from general statements (universal laws), to reach a conclusion in a particular case. Its purpose is to explain the content of the premises. If the premises are true, the conclusion can only be an indisputable truth, since the information was already contained in the premises, although implicitly. However, as the conclusion is drawn from the premises, it only confirms a truth; there is nothing new in the information generated from the analysis, although it is far from obvious. This type of reasoning is widely used in physics and mathematics, for example.

Deductive reasoning is considered to be limited as it does not increase the possibility of new discoveries. It just confirms what is already estimated as truth; it emerges from the demonstration of what was already implicit in the premises and, as a result, science is reduced only to knowledge originated in this path in search of truth, in the name of security and precision. Conversely, premises in inductive reasoning are particular observations and experiments (evidence) concerning conceptual generalizations (conclusions) that may or may not be true. The conclusions reached are truths not contained in the considered premises; they offer, therefore, novelty, creation and revolution.

Unlike the deductive method, which leads us to true conclusions if based on true premises, conclusions in the inductive method are only probable. However, it enables the extension of knowledge, because it is meant to broaden it, that is, lead to new discoveries. It lacks in precision, but if all the premises are true, the conclusion is probably true (but not necessarily true). Induction, in short, is the relationship between observational evidence and scientific generalization. Nowadays, science recognizes other abstractions of the possibilities of conducting scientific research, each appropriate to the type of investigation to be carried out, such as hypothetical-deductive, dialectical and phenomenological methods.

Bacon, in the second volume of the publication emblematically titled New Organon,[3,4] organized and proposed inductive reasoning, appropriate to the empirical quest for knowledge, distinguished by a clear commitment to observation and experimental proof as a condition for scientific fact. Martin (1926) stated that scientists before and at the time of Bacon reflected upon science also from the perspective of analysis and testing, but they

---

3 In Latin: Novum Organum.

4 The New Organon is the second part (out of six) of Advancement of Learning: *Book I – 'The Divisions of Science'*; *Book II – 'The New Organon or Direction Concerning the Interpretation of Nature'*; *Book III – 'The Phenomena of the Universe: or a Natural and Experimental History for the Foundation of Philosophy'*; *Book IV – 'The Ladder of the Intellect'*; *Book V – 'The Forerunners: or Anticipations of the New Philosophy'*; *Book VI – 'The New Philosophy: or Active Science'*.

draw conclusions by deductive reasoning, while the New Organon defended the formal validity of the inductive method in an innovative manner. Underlying the main concepts, his work clearly described the benefit of the progress of science for society and pointed to the state of permanent evolution of knowledge:

> Nothing for ourselves personally, but about what we are doing, we ask that men think of it not as an opinion but as a work and hold it for certain that we are laying the foundations not of a sect or of a dogma, but of human progress and empowerment. And then that they would give their own real interests a chance and put off the zeal and prejudice of beliefs and think of the common good; then, freed from obstacles and mistaken notions of the way, and equipped with our helps and assistance, we would ask them to undertake their share of the labors that remain[...] in fact, it is the end of unending error, and the right goal, and accepts the limitations of mortality and humanity, since it does not expect that the thing can be completely finished in the course of one lifetime, but provides for successors; and finally, that it seeks knowledge not (arrogantly) in the tiny cells of human intelligence but humbly in the wider world. (Bacon, 1605, in an edition included in Martin, 1926, for. 13)

Bacon is irretrievably linked to past and present; he is the link between the Renaissance[5] and the modern era. A contemporary of Galileo and Descartes, he was one of the first to understand the need to establish a new path for the quest for true knowledge. The fragment quoted above also illustrates his belief in science as the driving force behind the common good – a theme that will be addressed again later.

1.2 The Scientific Method of René Descartes (1596–1650)

Francis Bacon (1561–1626) and René Descartes (1596–1650) are considered to be the founders of modern science: Bacon as a pioneer of the experimental method and Descartes with the rigorous path in search of the truth. The work 'The New Organon', the second volume of the Advancement of Learning collection, is the focal point of the English philosopher's thought, while the French thinker wrote his fundamental principles in several different moments, and every single one of them confirms the defence of the unity of knowledge.

5 Period of European History between 1348 and 1648, approximately.

One example can be found in the work Rules for the Direction of the Mind, written in 1628, but only published posthumously in 1701 (in Latin):

> One also has to believe that all the sciences are so interconnected that it is much easier to learn them all together than to separate one from the others. Therefore, if one seriously wants to investigate the truth of things, then one should not select some particular science, for they are all interconnected and interdependent. Rather, one should simply think about increasing the natural light of reason, not in order that one might solve this or that scholastic difficulty, but in order that the intellect might show the will what has to be decided in the particular situations of life. Soon one will be amazed that one has made far greater progress than those who study particulars, as well as that one has achieved, not only all the things that the others desire, but also things higher than they could ever hope to expect. (Descartes 1998, p. 69)

The circumstances that motivated Descartes' intellectual production were the uncertainty about the principles that supported the pursuit of knowledge in that period and the method for conducting scientific research. So, he wrote 'Rules [...]', an unfinished set of three books of which only the first was published in full, with 12 rules, which are summarized below.

- Rule 1: on wisdom and science (the aim of studies should be to form solid and truthful judgments on any matters).
- Rule 2: on science and knowledge (the effort should only be put in what one can truly know).
- Rule 3: on intuition and deduction (on any subject to be investigated, one does not need to inquire about what other people think, or what we ourselves think, but all that matters is only what can be perceived with intuition or deduced with certainty, as there is no other way of acquiring knowledge).
- Rule 4: on method, mathematics and *mathesis universalis* (a method is needed to discover the truth).
- Rule 5: on the order and arrangement of things (the method consists entirely of the order and arrangement of objects towards which our mind must turn in the pursuit of the truth; thus, obscure propositions are reduced step by step to those which are simpler, and then we start the intuitive apprehension of the simple ones, ascending to the knowledge of all the others by similar steps).
- Rule 6: on simple natures and complex relations (separating what is simple from what is complex, isolating the simple and delimiting the interval that separates it from the others).

- Rule 7: on enumeration or induction (in order for one to achieve full knowledge, all matters which promote a purpose must be examined by a movement of thought that is continuous and nowhere interrupted).
- Rule 8: on the limits of human reason (if in the matters to be examined our understanding is sufficient to lead to knowledge, we must stop there in order not to make any attempt to examine what follows, thus sparing superfluous labour).
- Rule 9: on perspicacity of intuition (insignificant facts deserve full attention and must remain in contemplation for a long time until the truth is seen clearly and distinctly).
- Rule 10: on sagacity of deduction (the mind must be exercised to pursue only those inquiries for which the solution has already been found by others and must systematically traverse even the most insignificant human inventions).
- Rule 11: on capacity of inference (if, by intuitively recognizing a series of simple truths, we wish to draw any inference from them, it is useful to run them over in a continuous and uninterrupted act of thought, to reflect on their mutual relations and understand together, distinctly, several of these propositions, as far as possible, at the same time).
- Rule 12: on the power and operations of the natural intelligence (We should make use of understanding, imagination, sense and memory to have a distinct intuition of simple propositions and to compare the propositions).[6]

At this point, the following comment on the Rules is appropriate: some actually remain vigorous for the advancement of the scientific method; for example, Rules 11 and 12 inspire reflections on the integration of knowledge, on collaborative groups, on socialization, creation, interpretation, assimilation and accommodation of knowledge until reaching equilibrium, unlike Rule 3, which can virtually be disregarded today.

The Rules become, in fact, a prelude to Discourse on the Method,[7] of 1637 (in French), the monumental contribution of the mathematician, when he comes to the conclusion *cogito ergo sum*:

> Finally, considering that all the same thoughts which we have while awake can come to us while asleep without any one of them

6 *1 On wisdom and science; 2 On science and knowledge; 3 On intuition and deduction; 4 On method, mathematics and* mathesis universalis; *5 On the order and arrangement of things; 6 On simple natures an complex relations; 7 On enumeration or Induction; 8 On the limits of human reason; 9 On perspicacity of intuition; 10 On sagacity of deduction; 11 On capacity of inference; and 12 On the power and operations of the natural intelligence* (Descartes 1998, p. vi).

7 The full title was *Discourse on the Method of Rightly Conducting One's Reason and of Seeking Truth in the Sciences*, and it was written anonymously as a prologue to three other scientific texts.

> then being true, I resolved to pretend that everything that had ever entered my head was no more true than the illusions of my dreams. But immediately afterwards I noted that, while I was trying to think of all things being false in this way, it was necessarily the case that I, who was thinking them, had to be something; and observing this truth: I am thinking therefore I exist, was so secure and certain that it could not be shaken by any of the most extravagant suppositions of the sceptics, I judged that I could accept it without scruple, as the first principle of the philosophy I was seeking. (Descartes and Maclean 2006, p. 32)

His work, furthermore, summarizes four rules, which the philosopher deems sufficient to establish a method, that is, sufficient to sustain the logic that he revealed in his quest for truth. These are the rules: (i) evidence, never accepting anything as true without first judging it indisputably so, something that could not be possibly doubted, avoiding prejudice and premature conclusions; (ii) analysis, dividing all the difficulties under examination into as many parts as possible, and as many as necessary to solve them in the best way; (iii) order, conducting thoughts in a certain order, starting with objects that are the simplest and easiest to understand, and gradually ascending to the most complex; (iv) enumeration, the long chains of reasoning in the right order to deduce one thing from another; 'there can be nothing so remote that one cannot eventually reach it, nor so hidden that one cannot discover it' (p. 20).[8]

In the preface to the French translation of Principles of Philosophy (Descartes 1644) (*Principes de* philosophie, in 1647, 3 years after its launch in 1644, in Latin), he presented an image of knowledge relations in the form of a tree, an analogy that goes back to Ramon Llull's 1296 publication, entitled Tree of Science (Lulli 1515),[9] in which each science is represented by a tree with roots, trunk, branches, leaves and fruits. The roots represent the basic principles of each science; the trunk is the structure; the branches,

8 Descartes said: 'I venture to claim that the scrupulous observance of the few precepts I had chosen gave me such ease in unravelling all the questions [...] not only did I solve some which I had earlier judged very difficult, but [...] I was able to determine, even in respect of those questions which I had not solved, by what means and to what extent it was possible to solve them' (p. 21).

9 According to Norman (2020), none of Ramon Llull's books appear to have been published before the 15th century; the editions of *Arbor Scientiae*, with their famous woodcuts of Llull's trees of knowledge, began to appear in the early 16th century in the Lyon-printed edition in 1515.

the genera; the leaves, the species, and the fruits, the individual, his acts and his purposes. Llull's representation, in turn, is influenced by ancient Greek philosophers, especially the Aristotelian classifications. Matthews (1989) wrote:

> Descartes's life work became the creation of a systematic philosophy which would encompass all branches of knowledge. The system would be based on a few undeniable principles, and all knowledge would be deduced from them, so that metaphysics, physics, mathematics, morals, and politics would all coherent. Knowledge is an organic whole, in which all fields have the same method. Descartes repeatedly used the metaphor of a tree: 'Thus philosophy as a whole is like a tree whose roots are metaphysics, whose trunk is physics, and whose branches, which issue from this trunk, are all the other sciences'. This doctrine of a single, all-embracing method, is contrary to that of Aristotle, for whom the different fields of human knowledge all have their own subject matter and appropriate method. (pp. 87–88)

All of it is evidence of Descartes' unifying thought. According to Ariew (1992), 'From his earliest writings, the "Private Thoughts" for instance, we have Descartes' dream of a chain of sciences that would be no more difficult to retain than a series of numbers' (p. 111), or Rule 1: 'an explicit denial of the doctrine that the sciences should be distinguished by the diversity of their subjects, "all the sciences being in effect only human wisdom, which always remains one and identical to itself, however different are the objects to which it is applied"' (p. 111).

1.3 The Whole and Its Parts by Gottfried Wilhelm Leibniz (1646–1716)

Gottfried Wilhelm Leibniz was born in Germany in 1646, shortly before Descartes' death in 1650. The polymath stood out in several fields of knowledge, including philosophy, in which he recorded his perception of reality in several documents. For example, in: 'the true unity, unlike an abstract unity, is the one that encompasses an infinite variety or a world of diversities'[10] (Cardoso 2016, p. 20)[11] or 'I perceived that it is impossible to find the principles of a true unity in matter alone… since everything in it is only

10 Our translation.
11 In Letter to Arnauld, April 1687.

a collection or aggregation of parts to infinity'[12] (Cardoso 2016, p. 20)[13] or as put by Luca (2014, p. 18):

> Leibniz's stance must be well understood: far from any relativism, what the philosopher argues is that the unity of the world is represented on the basis of perspectives, that is, from divisions that we ourselves make to better understand the whole; and in this process of dividing/classifying the world to understand it, that is, systematically arranging disciplines, on the one hand, it can be indisputably better for acquiring clear and distinct knowledge, as it is a practical response to our needs (indexes, taxonomies, classification systems), but, on the other hand, what should not be lost sight of is that this entire body of particular sciences is one, continuous, uninterrupted, that is, it better achieves its natural flow by multiplying relations and connections that can be made between different types of knowledge.

Leibniz argues that a truth can be found in several realities, depending on one's particular point of view, and that the divisions of knowledge are arbitrary: 'they are not a consequence of the very nature of knowledge, but of our will'[14] (Hirata 2012, p. 24); in *Theophilus*[15]'s words: 'That division was a famous one even among the ancients [...] But the chief problem about that division of the sciences is that each of the branches appears to swallow the others' (Leibniz 2015, p. 258).

The mereology[16] presented in Leibniz's metaphysics originates in the work of Aristotle, whose part-whole axiom is known by some as an Aristotelian

12 Our translation.

13 In New System (1695).

14 Our translation.

15 *Theophilus* is the character from the book 'The New Essays on Human Understanding', which is an item-by-item rebuttal to John Locke's masterpiece 'An Essay on Human Understanding'. The two characters in the book are friends *Theophilus*, representing Leibniz's rationalism, and *Philalethes*, representing Locke's empiricism.

16 In philosophy and mathematics, mereology is the theory of relations between the parts and the whole. It studies the behaviour of part-to-whole relationships and part-to-part relations within a whole. It is rooted in the early days of philosophy; it began with the pre-Socratic philosophers and continued in the writings of Plato, Aristotle, and Boethius. Middle-age thinkers also addressed the issue, for example, Peter Abelard, Thomas Aquinas, Raymond Lull, John Duns Scotus, Walter Burley, William of Ockham and Jean Buridan. Contemporaneously, it is present in the works of Brentano and Husserl (Stanford Encyclopedia of Philosophy 2016).

principle (Atten 2017), owing to the words 'The relation of that which exceeds to that which is exceeded is numerically quite indefinite, [...] whereas that which exceeds, in relation to that which is exceeded, is "so much" plus something more' (Aristotle 1989, p. 1021a). According to Varzi (2016), the part-whole relationship represents a reflexive (since everything is part of itself), transitive (because any part of any part of a thing is itself part of that thing) and anti-symmetric relationship (since two distinct things cannot be part of each other).

For Burkhardt and Degen (1990), the part-whole relationship is an essential element of Leibniz's philosophy. In the work Dissertatio de arte combinatoria (Leibnüzio 1666),[17] Leibniz described the doctrine of the whole and its parts, drawing an important distinction: '[h]e separates the relation between the whole and its parts from the relation between the parts themselves, and he gives these different sorts of relations different names (p. 6).' In *Nouveaux Essais* (Leibniz 2015), the Latin expression '*omni et de nullo dictum*'[18] is used by Leibniz to illustrate that 'on the one hand not every containing thing is a whole and on the other hand every true whole is more than the parts, whereas the containing thing and that which is contained by it are in a certain way equal' (Burkhardt and Degen 1990, p. 6).

1.4 The Roots of Systemic Thinking in Johann Heinrich Lambert (1728–1777)

The ideas of Bacon, Descartes and Leibniz resulted in the actual separation between science and philosophy in the modern era. Empiricism and the inductive method took a leading role in the pursuit of scientific knowledge, and in the years to come, in the eighteenth century, the concept of complexity of science emerged with Johann Heinrich Lambert (1728–1777). Better known as physical and mathematical, Lambert's paradigm is about an approach to structuring complexity as a set of interrelated elements, described as various types of systems, such as systems of scientific knowledge, belief, cultures, religions, among many others (Hadorn 2008).

Your science, explained in Neues Organon (Lambert 1764), described what the scientific approach (both experimental and theoretical) should be like and constituted the foundation of systemic thinking. In the document, he wrote that human knowledge is partial and that to achieve it as a whole, one should adopt an interactive approach to the environment. Bullynck (2010) argues

17 It is the enlarged version of his doctoral thesis.
18 'The maxim of all and none'.

that 'nearly all items in Lambert's scientific output [...] testify of his systematic spirit and of his versatility in changing and adapting procedures to the circumstances'[19] (p. 67).

It is noteworthy that, like Bacon, Lambert used the word Organon in his publication on how to conduct scientific work, but he emphasized that it should be used and understood as originally conceived, that is, as a collection of tools to be used, combined and assembled according to the problem at hand. Lambert was concerned with the step-by-step process of the scientific experiment, with the *modus operandi*, with routines and subroutines, partly taken from other scientists, partly developed by himself (Bullynck 2010). And for that reason, he conceived an open approach of interacting and interdependent parts, 'mapping the ephemere geography of thinking, searching and finding. This is voiced in the conclusion to a long fragment on how to analytically transform experiments into a system.' (p. 67).

1.5 Kurt Gödel's Incompleteness Theorem (1906–1978)

Another important cornerstone for understanding the pursuit of knowledge is Kurt Gödel's Incompleteness Theorem (1906–1978). The scientist proved that mathematics is full of paradoxes, clearly demonstrating that there are true statements that cannot be proved, even if they are correct, and that the consistency of a system cannot be proved within the same system. These theorems revolutionized mathematics and expanded the foundations of the pursuit of knowledge, as the understanding of reality rose to another level.

In practice, the theorems led the scientific community to distinguish different levels of reality, to an increasingly deep and comprehensive knowledge and the certainty that all knowledge is equally important, overcoming the prejudiced hierarchy of knowledge. Nicolescu (1996a) noted that the Gödelian structure of reality levels implies that a complete theory can be built from a single perspective:

> A new Principle of Relativity emerges from the coexistence between complex plurality and open unity: no one level of Reality constitutes a privileged place from which one is able to understand all the other levels of Reality. A level of Reality is what it is because all the other levels exist at the same time. This Principle of Relativity is what originates a new perspective on religion, politics, art, education, and social life.

---

19 Full quote: *Nearly all items in Lambert's scientific output, especially those that are part of an ongoing series of investigations over the years, testify of his systematic spirit and of his versatility in changing and adapting procedures to the circumstances.*

And when our perspective on the world changes, the world changes. In the transdisciplinary vision, Reality is not only multidimensional, it is also multi-referential.

Unlike the disciplinary focus, multireferentiality and multidimensionality of knowledge result in the distinction of various levels of reality and, with this view, the approach shifts from the classical logic of disciplinarity to transdisciplinarity. Whatever the point of view, Gödel's propositions need to be considered in any modern study of knowledge; as assured by Nicolescu (1996b), it is 'at once the apex and the starting point of the decline of classical thought [...] [since] [t]he axiomatic dream is unraveled by the verdict of the holy of holies of classical thought ~ mathematical rigor' (p. 8).

The Incompleteness Theorem did not have repercussions in the 1930s and, in fact, only in the dawn of quantum mechanics did the Gödelian structure of the Reality levels turn out to be one of the most important contributions to science, unveiling the transdisciplinary approach. For this reason, Kurt Gödel is considered to be the father of transdisciplinary thought, which only started to be discussed four decades later, in the historic event in France.

1.6 Partial Considerations

Until the Renaissance, science was founded on *corpus aristotelicum*. Deductive logic was primarily taught in university curricula. The works of the Organon were discussed in the realm of philosophy, theology, medicine and law. However, from the mid-sixteenth century onwards, those who began to distinguish between the obsolescence of the prevailing theory of science and the need for a scientific methodology, emphasizing empiricism through the inductive method, emerged in the scientific-philosophical scenario of the time.

The transition to the modern period paved the way for the new science and, from then onwards, the foundations were laid for the quest for scientific and technological knowledge and the definitive separation between science and philosophy. It was followed by the scientific method, the bases of systemic thinking and the critique of the division of knowledge. The twentieth century posed new challenges to the intended unity of knowledge, including the distinction of various realities relative to the same element. The evolution of the way of perceiving and seeking true knowledge culminated in the Science Unity Movement which, in turn, impacted the study of the disciplinary relations of Jean Piaget, Andre Lichnerowicz and Erich Jantsch, which ultimately prompted the first definitions of transdisciplinarity.

Importantly, there are additional challenges to this equation, posed not only by scientific knowledge *stricto sensu*, but also by the intrinsic characteristics of the knowledge generated by technological research. Bunge (1985, p. 231), one of the pioneers of philosophical reflection on technological knowledge, defines technology as:

> the scientific study of the artificial or, equivalently, as R&D (research and development). If preferred, technology may be regarded as the field of knowledge concerned with designing artifacts and planning their realization, operation, adjustment, maintenance and monitoring in the light of scientific knowledge.

Cupani (2006) timely stressed that technology is different from applied science: 'many inventions did not originate from the deliberate application of scientific knowledge, nor were they designed by scientists. One of the best-known examples is the invention of the steam engine, which instead of being the result of scientific theories, contributed to the development of thermodynamics.'[20] Indeed, technology concerns the production of something new; it is a productive activity, related, according to Kroes (1989), to feasibility, reliability and efficiency, cost-effectiveness, and these are problems that do not usually affect basic science.

Technological knowledge, therefore, is an expression that brings together four typologies of knowledge related to practical usefulness. The first is tacit knowledge from domain experts, which is the foundation of knowledge management, related to subjective knowledge and acquired from individual experience. The second is the knowledge of practical rules, resulting from the definition of guidelines, conventions and standards, originating both from the combination and use of tacit knowledge and from the simplification of scientific knowledge. The third is scientific knowledge, resulting from scientific research, aimed at the development of engineering; it takes into account the conformation of society, economic aspects and limitations of the environment. It becomes high-tech or technology-intensive knowledge as the number of scientific aggregates is increased. And the fourth is knowledge of applied sciences, those whose objective is to apply existing knowledge to solve practical problems and, as a result, to promote the acquisition of new knowledge (Hansson 2013).

---

20 *numerosas invenciones no se originaron en la aplicación deliberada de conocimientos científicos ni fueron realizadas por científicos. Uno de los casos más conocidos es la invención de la máquina de vapor, que en vez de ser el resultado de teorías científicas contribuyó a desarrollar la termodinâmica.*

The content of this chapter intended to present the axes for the definition of unity of knowledge,[21] a fundamental concept of transdisciplinarity. The ideas presented are translated into the following words: unity of knowledge is the concept that recognizes that we must seek the systematic unity and integrity of knowledge, since the very nature of logic and reason lies in the activity of integrating propositions under increasingly general principles, in order to systematize, unify and complete the knowledge acquired through real understanding. It means to unify it more and more in the light of the idea of a whole of knowledge; this way, the interacting parts are displayed according to convenience. As a result, unifying knowledge means recognizing the multiplicity of theories, in several fields, which must complete one another to create increasingly integrated and efficient explanatory models for describing and accounting for the world. However, it is worth noting that partially unified knowledge can progress in a law-like manner to a single final theory, which can never really be reached, because the task of knowledge is infinite; there will always be more to understand and deeper explanations to give.

In other words, unity of knowledge refers to the fact that conditioned knowledge can only achieve its integrity through unconditioned knowledge; the latter is not just a contingent aggregate, but an indispensable system for the complete identity of conditioned knowledge.

21 The definition adopted here was also based on the Principle of the Systematic Unity of Reason by Immannuel Kant (1724–1804), on the Theory of Objective Validity of neo-Kantians (1870–World War I), on Karl Ludwig von Bertalanffy's General Systems Theory (1901–1972) and on Consilience, by Edward O. Wilson (1998), whose content will be explored in other works.

Chapter 2

TRANSDISCIPLINARITY

2.1 Benchmarks

2.1.1 The Disciplinary Relations of Jean Piaget and Andre Lichnerowicz in the 1970s

The contribution of Jean Piaget and André Lichnerowicz to transdisciplinarity began with the 1970 event[1] and extended for some time until each one of them embraced their main areas of interest, for which they became known. Biologist Piaget dedicated his whole life to studying the process of knowledge acquisition, especially by children. Lichnerowicz was dedicated to the study of differential geometry and was recognized for his contribution in the field, for example, he chaired the Lichnerowicz Commission, established to analyse the pedagogical project of teaching mathematics.

In the early 1970s, they were both involved in teaching and learning issues of science, and Piaget, in particular, was aware of initiatives in the scientific community for the debate on disciplinarity, including the Unity of Science movement from the first half of the twentieth century (1922–1936), founded by a group of scientists and philosophers who met regularly at the University of Vienna (hence it was called the Vienna Circle, also known as the 'Ernst Mach Society'). The movement argued that there should be a unitary set of physical premises from which the regularities of all reality could be derived.[2]

During the 1970 event, this vision was contained in the reflections of his work '*L'épistémologie des relations interdisciplinaires*',[3] as well as in the necessary distinctions

1 Whose proceedings were published in 1972.

2 Members included Friedrich Waismann, Gustav Bergmann, Hans Hahn, Herbert Feigl, Karl Menger, Ludwig von Bertalanffy, Marcel Natkin, Olga Hahn-Neurath, Otto Neurath, Philipp Frank, Richard von Mises, Rose Rand, Rudolf Carnap, Theodor Radakovic, Tscha Hung, Victor Kraft, Hans Reichenbach, Kurt Gödel, Carl Hempel, Alfred Tarski, W. V. Quine and A. J. Ayer (Stanford Encyclopedia of Philosophy 2017b).

3 The epistemology of interdisciplinary relations.

between interdisciplinarity and multidisciplinarity, the results of which led him, in fact, to coin the term and the first concept of transdisciplinarity:

> Finally, we hope to see succeeding to the stage of interdisciplinary relations a superior stage, which should be 'transdisciplinary', i.e., which will not be limited to recognize the interactions and or reciprocities between the specialized researches, but which will locate these links inside a total system without stable boundaries between the disciplines. (Piaget 1972, p. 144)

Throughout the Conference, it is evident that the educator saw transdisciplinarity as a new form of disciplinary relations, more integrative than interdisciplinarity, going beyond and even being the result of interdisciplinarity. Nicolescu (2006) pointed out that 'this description is vague, but has the merit of pointing to a new space of knowledge "without stable boundaries between the disciplines" ' (p. 1).

At the end of his work, however, he attributed to André Lichnerowicz a deeper analysis of the concept of transdisciplinarity: 'As for specifying what such a concept might encompass, it would obviously be a matter of a general theory of systems or structures. [...] It is up to the mathematician to tell us more and Lichnerowicz will enlighten us about this future'[4] (Piaget 1972, p. 171).

Lichnerowicz's approach is, as expected, mathematical, as this is his background. He adopted the concept of isomorphism to explain transdisciplinarity. In mathematics, isomorphism involves recognizing the phenomena of one object in another, that is, if two objects are isomorphic, then any property that is preserved by one isomorphism is also true for the other object. This function can be used to investigate problems from an unknown field based on another field whose problems are clarified. Therefore, for him, 'transdisciplinarity, consists in treating by the same mathematical model (isomorphism) disciplines of a very different nature, but according to the same laws'[5] (Lichnerowicz 1980, p. 22).[6]

4 *Quant à préciser ce que peut recouvrir un tel concept, il s'girait évi- demment d'une théorie générale des systèmes ou des structures [...] c'est au mathématicien à nous en dire davantage et Lichnerowicz nous éclairera sur cet avenir.*

5 *« transdisciplinarité », elle consiste à traiter par le même modèle mathématique (isomorphisme) des disciplines de nature fort différente, mais obéissant aux mêmes lois.*

6 His 1980 work resumed his 1973 publication: Lichnerowicz, A. (1973). Mathématique, structuralisme et transdisciplinarité. In Rheinisch-Westfälischen Akademie der Wissenschaften (Eds.). *Natur-, Ingenieur-und Wirtschaftswissenschaften.* Wiesbaden (Germany): VS Verlag für Sozialwissenschaften.

For him, the result of this understanding is that regardless of the field of knowledge, theoretical activity is homogeneous throughout science and technology, which 'assumes and imposes a certain transdisciplinarity'[7] (p. 31). In a simplified definition, the mathematician saw transdisciplinarity as an 'angle of vision that goes far beyond artificially limited disciplines as subjects of knowledge'[8] (p. 31) and clearly expresses his concern with higher education, which ignores this condition:

> All over the world, our present universities form, it seems to me, a very large proportion of specialists in predetermined disciplines, therefore artificially limited, while a large part of social activities, such as the development of science itself, require men capable of having both a broader vision and in-depth focus on new problems or projects, transgressing the historical limits of the disciplines. These are the men we also need to form.[9] (Lichnerowicz 1980, p. 32)

Piaget's and Lichnerowicz's definitions lie in the epistemological field. The third participant in the event that dealt with the theme, Erich Jantsch, has another approach.

2.1.2 The Social Purpose of Erich Jantsch in the 1970s

Erich Jantsch (1929–1980) defined transdisciplinarity from another perspective. He defined it on the basis of coordinate systems for a common purpose (Jantsch 1972). He suggested that knowledge should be organized in objective-oriented hierarchical systems, and by creating the design of such structures, he introduced the concept of transdisciplinarity.

He realized that multilevel and multiobjective systems, based on transdisciplinary management, would be the ideal structures to fully achieve scientific knowledge. From this point of view, he attributed to transdisciplinarity the path to scientific development. Klein (2009) claimed

---

7 *qui supposent et imposent une certaine transdisciplinarité.*

8 *Angle de vue dépassant largement les disciplines artificiellement bornées em tant que matières des connaissance.*

9 *A travers le monde nos universités présentes forment, me semblet-il, une proportion trop grande de spécialistes de disciplines pré- déterminées, donc artificiellement bornées, alors qu'une grande partie des activités sociales, comme le développement même de la science, demandent des hommes capables à la fois d'un angle de vue beaucoup plus large et d'une focalisation en profondeur sur des problèmes ou des projets nouveaux, transgressant les frontières historiques des disciplines. Ce sont ces hommes qu'il nous faut aussi former.*

that, of the pioneering models, Jantsch's became the most influential. It was adapted as a conceptual framework in several fields by structuring his concept of transdisciplinarity on the basis of the general systems theory and the organization theory. And 'a new relationship between science and society echoed in critiques of traditional notions of "objectivity" and "progress"' (Klein 2009, n.p.). For Jantsch, in the words of Klein (2009):

> The effects would be pervasive. New types of institutions would be needed and a new form of education capable of fostering the capacity for judgment in complex and dynamically changing situations. In science, technology and industry, long-range thinking would replace short-range thinking. In cities and the environment, negative effects of technology would be reversed, and a systems approach would replace linear modes of problem solving. The university would also gain a new purpose. (n.p.)

For Nicolescu (2006), Jantsch's historical merit was to highlight the need for an axiomatic approach (he imagined disciplines and interdisciplines coordinated by a generalized axiomatics) to transdisciplinarity and also to the introduction of new values in this field of knowledge.

2.1.3 Mode 2 Knowledge Production by Michael Gibbons, Camille Limoges, Helga Nowotny, Simon Schwartzman, Peter Scott and Martin Trow in 1994

In the study led by Michael Gibbons in 1994, the authors came to the conclusion that many problems are not within a disciplinary framework and a traditional way of pursuing knowledge, which he called Mode 1, characterized by homogeneity and hierarchy. Rather, the problems that give rise to learning are defined and resolved in a context governed by the interests of a specific community. The new mode is performed in a non-hierarchical, heterogeneous way and involves the close interaction of many actors throughout the knowledge production process. As a result, knowledge production is becoming more socially responsible and more reflective, and it affects science at deeper levels of transformation. This new form of knowledge production, called Mode 2,

> results from a broader range of considerations. Such knowledge is intended to be useful to someone whether in industry or government, or society more generally and this imperative is present from the beginning.

> Knowledge is always produced under an aspect of continuous negotiation and it will not be produced unless and until the interests of the various actors are included. Such is the context of application. [...] knowledge production in Mode 2 is the outcome of a process in which supply and demand factors can be said to operate, but the sources of supply are increasingly diverse, as are the demands for differentiated forms of specialist knowledge. (Gibbons et al. 1994, p. 4)

To qualify Mode 2's specific form of transdisciplinary knowledge production, the authors identified four distinct characteristics of transdisciplinarity. First, it 'develops a distinct but evolving framework to guide problem-solving efforts [...] generated and sustained in the context of application and not developed first and then applied to that context later by a different group of practitioners' (p. 5). The solution, therefore, will not only arise from the application of existing knowledge that will be integrated into the structure, but also from the experience and creativity of the application context itself. After reaching a consensus, the knowledge resulting from the solution can hardly be reduced to disciplinary parts.

According to the solution, it comprises both theoretical and empirical resources, since 'transdisciplinary knowledge develops its own distinct theoretical structures, research methods and modes of practice' (p. 5) whose result may not necessarily be disciplinary knowledge. Indeed, the cumulative results of an effort to offer a transdisciplinary solution to a problem, even if originated in a specific situation, can certainly go in different directions beyond the academy and the application context, including the emergence of contents that may not be located on the current disciplines available.

Third, the communication of results begins during the knowledge production process itself. Communication is now destined to the new contexts of the problem. Mode 2 highlights the fact that both the problems and the team responsible for the solution can be highly transitory and that is why the generated information needs to be safeguarded, especially in information networks that 'tend to persist and the knowledge contained in them is available to enter into further configurations' (p. 5).

Fourth, 'transdisciplinarity is dynamic. It is problem solving capability on the move' (p. 5), as Mode 2 is characterized 'by the ever closer interaction of knowledge production with a succession of problem contexts'. Another aspect is that the results are most often outside a specific disciplinary context, as knowledge produced in this way does not fit into any of the disciplines that contributed to its solution.

2.1.4 Transdisciplinarity Axioms of Basarab Nicolescu 1996

Gibbons' 1994 contribution coincided with the release of the Transdisciplinarity Charter in the same year during the First World Congress on Transdisciplinarity in Portugal.[10] The text document, written by Lima de Freitas, Edgar Morin and Basarab Nicolescu, was signed by 62 participants from 14 countries and provided input for understanding the epistemology of transdisciplinarity. Some excerpts addressed disciplinary relations, the recognition of the multidimensionality of reality and the need to contextualize knowledge.

Despite the theme being studied, the Charter of Transdisciplinarity and Mode 2 Knowledge Production remained as the last fundamental references on the subject until 1996, when physicist Basarab Nicolescu published the Manifesto of Transdisciplinarity (Nicolescu 1996b). In the document, he discussed the need for a methodology for transdisciplinarity, as a way to ensure that it can advance and effectively contribute in the various spheres of its potential. To this end, his choice was to point out three fundamental axioms for the methodology of transdisciplinarity: the ontology axiom, the logic axiom and the complexity axiom (Nicolescu 1996b).

2.1.4.1 The Ontology Axiom

It is the statement that deals with the identity of transdisciplinarity, based on different levels of reality and, correspondingly, different levels of perception. The axiom draws on the Principle of Relativity, which asserts that 'no level of Reality is a privileged place from which one is able to understand all other levels of Reality' (Nicolescu 1996b, p. 11). That is, the approach is not hierarchical; there is no fundamental level of reality; all realities exist at the same time and all of them matter.

As a result, each level is characterized by its incompleteness (reaffirming the Gödelian reality), that is, 'the laws governing this level are just a part of the totality of laws governing all levels. And even the totality of laws does not exhaust the entire Reality' (p. 11). Another immediate consequence of reality in a structure of levels implies that knowledge needs an open structure to advance and complete itself.

Incidentally, Nicolescu (2006) explained that the levels of reality were also addressed by Werner Heisenberg in 1942,[11] a decade after Kurt

---

10 Held at the Convento de Arrábida, Portugal, from 2 November 1994 to 6 November 1994.

11 The preface by Konrad Kleinknecht (President of the Heisenberg Society) in the first online version of the 2019 work reads: 'Werner Heisenberg, the inventor of quantum

Gödel's Theorem, which disclosed the various levels of Reality. In fact, Reality and its Order (Heisenberg 1989) was only published nearly 50 years later, in 1989, after Heisenberg's death, with the following understanding of reality: 'the continuous fluctuation of the experience as captured by consciousness. In that sense, it can never be identified to a closed system [...] One can never reach an exact and complete portrait of reality' (Heisenberg 1942 as quoted in Nicolescu 2006, p. 15).

2.1.4.2 The Logic Axiom

When one understands the different levels of reality and perception, classical logic gives way to the logic of the included third term. Binary logic consists of two values –true or false – based on the impossibility of a third possibility. That is, 'A is A' or 'not-A is not-A' and there is no third possibility, that is, that A is at the same time not-A. This situation is consistent with human reasoning.

The possibility of the existence of the third term T, which is at the same time A and non-A, is also a structuring element of transdisciplinarity. Thus, the logic of the included third term[12] allows different perspectives, for example, these opposites are not eliminated; rather, they coexist. In this condition, the passage from one level of reality to another is ensured by the logic of the included middle, which is actually the included third.

> In other words, the action of the logic of the included middle on the different levels of Reality[13] induces an open structure of the unity of levels of Reality. This structure has considerable consequences for the theory of knowledge because it implies the impossibility of a self-enclosed complete theory. Knowledge is forever open. (Nicolescu 2006, p. 18)

---

mechanics and one of the most important scientists of the 20th century, wrote this essay during the 1941/1942 war years. [...] It was published in 1989 by Piper Verlag'. (*Werner Heisenberg, der Erfnder der Quantenmechanik und einer der bedeutendsten Wissenschaftler des 20. Jahrhunderts, schrieb diesen Essay in den Kriegsjahren 1941/1942. [...] Er wurde 1989 im Piper Verlag veröfentlicht.*)

12 Advances in this area are due to discoveries in quantum physics, in the Quantum Superposition Principle. Continuing the discovery, philosopher Stéphane Lupasco (1900–1988) developed a non-classical logic emphasizing the balance between the poles of a contradiction, between the dynamics of heterogeneity and the dynamics of homogeneity.

13 Reality is always written with a capital letter R in Nicolescu's work and it will also be spelled this way in the present book.

2.1.4.3 The Complexity Axiom, Especially Edgar Morin's thought 1991

There are several theories of complexity, but Edgar Morin's is compatible with the levels of Reality described in the ontology axiom, referred to as vertical complexity.[14] It is an approach that avoids fragmentation through systemic reasoning, integrating the antagonistic poles of the contradiction of reality, multiplicity, randomness and uncertainty.

To fully understand the complexity axiom – or universal interdependence – one needs to understand its three underlying principles. The first, the Dialogic Principle, brings the premise that there are two logics that are at the same time complementary and antagonistic; there are opposing forces at work that are vital for the system to function, and they collaborate and produce organization and complexity. The Dialogic Principle allows us to maintain duality within unity (Morin 1990, here in a 2005 edition).

The second, the Principle of Organizational Recursion, deals with processes that are at the same time products and producers, that is, everything that is produced returns to what produces it in a recursive cycle of cause and effect, contrary to the idea of linearity. The example given by Morin (1990, here in a 2005 edition) is that 'individuals produce the society that produces individuals'[15] (p. 100).

The third, the Holographic Principle,[16] reinforces the relation between the whole and its parts. Morin (1990, aqui em edição de 2005) explained that not only is the part in the whole, but the whole is in the part. The very holographic idea of knowledge of the parts by the whole and the whole by the parts is related to the recursive idea: 'the whole is in the part that is in the whole'[17] (p. 101).

> According to Morin, the sum of the knowledge of the parts is not enough for one to know the properties of the whole, since the whole is greater than the sum of its parts. Furthermore, when one takes the whole, one cannot the richness of the qualities of the parts, as they are inhibited and virtualised, prevented from expressing themselves in their fullness. Hence the whole is less than the sum of its parts. The relations of the parts to the whole are dynamic, so the whole is both smaller and larger than the sum of its parts. (Santos 2008, p. 73)

---

14 Horizontal complexity refers to a single level of Reality, while in transversal complexity, different levels of organization intersect into a single level of Reality.

15 *les individus produisent la société qui produit les individus.*

16 Also translated as the Holographic Principle.

17 *Le tout est dans la partie qui est dans le tout.*

2.2 Meanings

2.2.1 Between Society and the University of Joseph Kockelmans

The work *Interdisciplinarity and higher education*, published in 1979 by Joseph J. Kockelmans, is fertile in discussing the various perspectives of transdisciplinarity. The chapter he wrote revolves around the condition about the transdisciplinary work, which is 'impossible without a philosophical reflection on man and society'[18] (p. 145), which the philosopher perceives in four distinct views.

The first view concerns the disintegration of the unity of knowledge, which started with the fragmentation arising from specialization. Although necessary, further specialization of each discipline resulted in the 'dehumanization of both man and his environment (p. 147)', since specialization itself must arise from a genuine concern for the whole. Thus, transdisciplinary efforts focusing on the whole are necessary: 'the transdisciplinary concern for the unity of our world must be there first, and from its specialization should flow' (p. 147).

The second view, derived from the first, is related to the positive contribution that philosophy can make to the transdisciplinary unification of sciences, since the philosophical dimension is present in all theoretical efforts and, therefore, also in the concern with the unity of knowledge. Furthermore, Kockelmans (1979) pointed out that scientists are so deeply engaged in research that they forget that they must first be concerned with the whole, and only a collective effort could re-establish priorities.

The third group, for the author, advocates the social relevance of higher education, and it is the first objective of transdisciplinarity to re-establish the contact between university and society. Considering that relevant social problems cannot be addressed by interdisciplinary efforts alone, Kockelmans (1979) wrote:

> The difference between crossdisciplinarity and transdisciplinarity consists in the fact that crossdisciplinary work is primarily concerned with finding a reasonable solution for the problems that are so investigated, whereas transdisciplinary work is concerned primarily with the development of an overarching framework from which the selected problems and other similar problems should be approached. (p. 128)

---

18 *Original quote: that genuine transdisciplinary work is impossible without a philosophical reflection on man and society.*

Broadly, the concepts of interdisciplinarity and transdisciplinarity converge to overcome the loss of the unity of knowledge; however, they maintain significant conceptual and methodological differences. For Weingart (2010), interdisciplinarity is the production of knowledge that crosses disciplinary boundaries, and transdisciplinarity is the growing effort to make knowledge products more relevant to non-academic actors. For Jantsch (1972), transdisciplinarity is 'purpose-oriented interdisciplinarity', a unifying principle for integration of knowledge, a process by which members of different teams work together to apply and integrate knowledge in society, in science and in technology.

For Stokols et al. (2010), transdisciplinarity not only integrates but also transcends disciplinary perspectives into shared conceptual and methodological frameworks developed for a purpose. They reinforced the point of view that the inclusion of non-academic collaborators in transdisciplinarity outweighs the aspects of synergy and integration, and they highlighted that, as far as interdisciplinarity is concerned, the results are more difficult to achieve, owing to the increasingly complex problems and transcendent goals.

And for the last group, Kockelmans' fourth vision, science has become the foundation of the life we live; it teaches the truth and is the most powerful instrument to change the world, for better or for worse. Therefore, transdisciplinarity emphasizes the correct conception of science, that is, genuine science must have critically analysed assumptions to guarantee 'that through the sciences man can provide for himself a position within the cosmos that is at the same time rational, critical, and humane'[19] (p. 150).

2.2.2 The Transcending Continuum by Christian Pohl

By analysing current definitions of transdisciplinarity, Pohl (2010) recognized two common patterns. In the first, he realized that definitions generally propose a progression from multidisciplinarity, through interdisciplinarity to transdisciplinarity, 'to transcend their separate conceptual, theoretical, and methodological orientations in order to develop a shared approach to the research, building on a common conceptual framework' (Rosenfield 1992 as quoted in Pohl 2010).

It is worth emphasizing the use of the term 'transcend' in the definition above. Klein (2010) highlighted it as underlying the essence of transdisciplinarity and,

19 *That through the sciences man can provide for himself a position within the cosmos that is at the same time rational, critical, and humane.*

in effect, its impetus. In the words of the emeritus professor and prolific author and researcher on the subject, transdisciplinarity is an 'integrative curriculum design driven by the keyword "transcending" ' (p. 24). The author also noted that the concept has been extended in some countries to a transdisciplinary science, as a 'collaborative form of "transcendent interdisciplinary research" that creates new methodological and theoretical frameworks' (p. 25).

Many other authors have also used the term to express the idea of transdisciplinarity, including Raymond C. Miller, who expressed his understanding with the same foundation as Erich Jantsch. In the synthesis of his definition, he used both the image of articulated structures, such as the pioneering thinker, and the transcendent aspiration: 'articulated conceptual frameworks which claim to transcend the narrow scope of disciplinary world views' (Miller 1982, p. 11).

The second pattern revealed that there are four possible situations for characterization of transdisciplinarity: (1) looking at relevant social issues, (2) going beyond disciplinary paradigms, (3) conducting participatory research and (4) searching for a unity of knowledge. The presence or absence of these situations led Pohl (2010) to organize them into three groups, with questions 1 and 2 being present in all three. Thus, Group A's definitions only contain these questions (1 and 2), Group B includes question 3 (1, 2, and 3) and Group C contains question 4 (1, 2, and 4).

In Group A, understanding transcends and integrates disciplinary paradigms in order to address socially relevant issues. Jantsch's (1972) definition advocates these socially relevant issues. The pioneer, followed by others, highlights that the disciplinary approach is increasingly shifting away from social problems and concerns.

Group B expands the concept to include the participation of non-academic actors described in the participatory research on Mode 2 knowledge production, and it was followed by Kötter and Balsiger (1999), Scholz and Tietje (2002), Lawrence (2004) and Mobjork (2010) (Pohl 2010). It should be noted that in the United States, this Group, which represents the participatory approach, is called action research or 'transcendent interdisciplinary research' (p. 194).

In fact, participatory research has countless possibilities and terminologies, and this book provides a brief presentation of action research and case study, understood as applied research, aimed at generating knowledge to solve specific problems and whose scientific method is empiricism, which considers experience as the basis of true knowledge and whose generalizations derive from the observation of reality on the basis of particular cases.

The term action research was coined by Kurt Lewin in 1946 as mediation research on critical social issues, through the researcher's action upon or in the social system, which compares the conditions and effects of various forms of action, resulting in knowledge after completion (Lewin 1946). These are major characteristics of action research: (i) it is future-oriented; (ii) it is participatory; (iii) it implies the development of a system that can guarantee all phases of research (diagnosis, planning, action, evaluation and learning); (iv) it generates action-based theory; (v) it recognizes that research objectives, problem and method must be generated from the research process itself and (vi) it is situational, as the relations between people, events and things are the result of how the relevant actors have defined the situation.

In turn, the case study is characterized by the in-depth study of a specific complex phenomenon, analysed from different perspectives, which reveals elements that can be found in other cases, for example, the principle of isomorphism, adopted by Lichnerowicz (1980) to explain transdisciplinarity. It is, in fact, a strategy that uses numerous methodologies to provide insights into the various connected dimensions of a specific social circumstance, using different techniques at different empirical scales and data integration by researchers from different technical and knowledge areas (Fidel 1984).

Back to the characterization of transdisciplinarity, Group C seeks a unity of knowledge, with the aim of reorganizing academic knowledge to make it useful for addressing relevant social issues, in a perspective that goes beyond all disciplines. Pohl (2010) points Nicolescu (1996b) and Ramadier (2004) as representatives of that Group.

2.2.3 The Abundance of the Concept for Julie Klein

Julie Klein (2013) described five main groups of meaning of transdisciplinarity: those related to conceptions of interdisciplinarity, those related to the unity of knowledge, to the alignments of participatory and collaborative research, to new forms of knowledge contained in transdisciplinarity, and about its transgressive aspect, which challenges the existing structure of knowledge, culture and education. The author stressed that, despite the differences in approach, these groups communicate with one another, in the construction of a 'structured plurality of definitions', an expression used by Pohl (2010).

2.2.3.1 From Disciplinarity to Interdisciplinarity

In the first approach, the concept goes from disciplinarity to the restrictions inherent in interdisciplinarity and emphasizes that the prefix *trans* leads to the idea of transcendence, synthesis and integration. To demonstrate the holistic aspect of the concept, the author refers to the Organization for Economic Cooperation and Development (OECD) definition of transdisciplinarity as a 'common system of axioms that transcends the narrow scope of individual disciplines through an overarching synthesis' (Klein 2013, p. 190), in contrast to the concept of interdisciplinarity also proposed by OECD (although it is a broad one): 'any form of interaction from simple borrowing of a method to a new paradigm for research and education' (p. 190).

2.2.3.2 From Unity to Complexity

Out of the group of definitions that reflect the discourse of the unity of knowledge, the associated concepts are, above all, uncertainty, diversity, non-linearity, multidimensionality, heterogeneity and relationality, supplanting the concepts of certainty, universality, simplicity, linearity and unidimensionality. The dichotomies presented above are derived from the problems created by the fragmentation of knowledge resulting from the expansion of the number of disciplinary specialties, which ignores the logic axiom of transdisciplinarity, the included third term, in which all realities coexist and all of them matter. The derived production of knowledge readily attributes the values of interaction, intersection and interdependence to the new logic of unifying approaches.

2.2.3.3 Participation and Collaboration

The approach that encourages collaborative research and participatory research alludes to cooperation, partnership and mutual learning. Collaborative research is the result of scientific collaboration and has become increasingly common as it has the potential to solve problems that are more complex and promote new political, economic and social agendas. 'Scientific collaboration is also referred to as research collaboration' (Sonnenwald 2007, p. 644) and can be defined 'as interaction taking place within a social context among two or more scientists that facilitates the sharing of meaning and completion of tasks with respect to a mutually shared, superordinate goal' (p. 645). The phenomenon of collaboration is not new, but in recent years, it has been intensified in all knowledge areas. Wray (2002) pointed out five reasons for the rise of collaborative research: it (i) increases the quality of research; (ii) increases explanatory coherence, especially when it involves scientists from different disciplines, favouring 'conceptual combinations that establish new theoretical frameworks'; (iii) reduces the possibility of omissions or forgetfulness of previous findings; (iv) accelerates the achievement of results and (v) plays an important role in the education of young scientists, in collaborations that bring together masters and apprentices.

Participatory research, on the other hand, was evident in the environmental and sustainability research carried out in the 1980s in Germany and Switzerland. It refers to a strategy that emphasizes and explores local knowledge, with the participation and perception of actors in that reality. In participation, the emphasis is on a process of 'reflection and action, carried out with and by local people rather than on them' (Cornwall and Jewkes 1995, p. 1667) and whose main difference with conventional research lies in the location of power in the research process, which can be abstracted in the expression 'knowledge for action' instead of 'knowledge for understanding' (p. 1667).

Cornwall and Jewkes (1995) argue that this modality involves various degrees of participation, which can be organized into four possibilities: (i) contractual, when people are hired by projects in order to take part in research; (ii) consultative, when people in the context are consulted about their opinions by researchers before interventions are made; (iii) collaborative, when researchers and people from the context work together on projects designed, initiated and managed by researchers and (iv) collegiate, when researchers and people from the context 'work together as colleagues with different skills to offer, in a process of mutual learning where local people have control over the process' (p. 1669).

There are four relevant questions that lead to participatory research, structurally guided by the common good: (i) socially relevant issues, (ii) the need for transcendence and integration of disciplinary paradigms, (iii) the convenience of participatory research and (iv) the pursuit of the unity of knowledge. The underlying premise is that social problems, derived from an increasingly complex and interdependent society, are not isolated in academic disciplines; on the contrary, 'they are emergent phenomena with non-linear dynamics, uncertainties, high political stakes in decision making, and divergent values and factual knowledge' (Klein 2013, p. 193), which require academic integration for solutions arising from science (*Science of Team Science*[20]).

2.2.3.4 Forms of Knowledge of Transdisciplinary Research

The penultimate group highlights the forms of knowledge, especially the interdependence of system knowledge: target knowledge and transformation knowledge, which are also the principles for the transdisciplinary research project presented by Hadorn et al. (2008). For them, transdisciplinarity

20 SciTS.

is the necessary integration throughout the knowledge system (knowledge of the system) in order to deal with the uncertainties that also result from the lack of empirical or theoretical knowledge about a problem. Target knowledge addresses the multiplicity of social situations that create a need to specifically define a research problem and its stakeholders in society and science, considering that their respective participation should lead to the development of knowledge and practices that promote the common good (e.g., also pointed out by the third group of definitions identified by Kockelmans (1979). Transformation knowledge is the technologies, regulations, practices and relationships that are present in the context of the project. Existing infrastructure, power relations and cultural preferences must be considered as constituents of transformation knowledge.

Here it is worth highlighting the setting of the transformation knowledge infrastructure with the announced knowledge industry. The relatively recent analysis by Ghassib (2012) of the knowledge industry, as with any other industry, is based on: (i) production sites; (ii) producers; (iii) production instruments; (iv) raw materials; (v) production methods and (vi) production outputs. These are the foundations that provide a glimpse of the infrastructure needed to sustain transformation knowledge. It is noteworthy that Fritz Machlup, in his 1962 work (Machlup 1962), already demonstrated the emergence of the knowledge industry by mapping its production and distribution in some sectors of the economy in the United States.

Also in this group is the awareness of co-production of knowledge, 'a new social distribution of knowledge is occurring as a wider range of organisations and stakeholders are involved, including NGOs, private firms, and governmental agencies' (p. 196). Or as pointed out in the extension of the work of Gibbons et al. (1994) (Nowotny et al. 2013), the

> contextualization of problems requires participation in the agora of public debate, incorporating the discourse of democracy. When lay perspective and alternative knowledges are recognized, a shift occurs from solely 'reliable scientific knowledge' to inclusion of 'socially robust knowledge', dismantling the academic expert/non-academic lay dichotomy. (Klein 2013, p. 196)

Social robustness is a term widely used to characterize transdisciplinary research – the quality and application of the results achieved, the relevance, effectiveness and accessibility of intervention in the social system. Nowotny et al. (2013) stressed that robustness is not an absolute concept, nor is it a relative one. It is a relational concept; it depends on the following considerations: (i) it can only be judged in specific contexts; (ii) it describes a process that at a timely moment can reach a certain degree of stability; (iii) there is a subtle distinction between robustness of knowledge and its acceptability (by individuals, groups or societies); (iv) it is produced when it is incorporated and enhanced by social knowledge and (v) socially robust knowledge has a strongly empirical dimension.

It is worth adding that socially robust outcomes can include mutual learning, trust-building between stakeholders, establishment of new relationships, advancement of knowledge and greater ability to work as a team and articulate common goals (Polk 2011).

2.2.3.5 The Transgressive Imperative

The last group of concepts reveals the criticism of the existing structure of knowledge, culture and education and the need for transformation. In addition to issues related to struggles for social change, issues of culture regarding the limits of class, gender, race, ethnicity and other identities and human rights, at whose core are the foundations of transdisciplinarity (as well as movements that reject disciplinarity), the pattern here refers to the distance (and ultimately separation) between tradition and science, West and non-West, theory and practice, and other dichotomies that ignore the varied forms of knowledge. The challenge faced by transdisciplinarity is the capacity to overcome the divisions that affect research, practice and learning, intensifying the awareness of heterogeneity, incorporating forms of knowledge that were previously excluded and, thus, increasing the relationality of knowledge.

2.3 Partial Considerations

The concept of transdisciplinarity is quite broad. It was first used in the 1970s in the work of Jean Piaget and André Lichnerowicz while they were investigating disciplinary relations, seeking to advance knowledge beyond the discipline. On the same occasion, Erich Jantsch expressed the need to organize knowledge on the basis of hierarchical systems, in whose intersection transdisciplinarity would lie for the purpose of coordination. In this position, the concept of transdisciplinarity, according to the author, advances towards solidarity and common purpose. The foundations of transdisciplinarity are widespread and its origins date back to 1979 in Joseph Kockelmans' theoretical propositions.

For a long time, these were the only milestones of the concept of transdisciplinary, but in 1994, Michael Gibbons and his group developed Mode 2 knowledge production. They stressed that problems do not lie within a disciplinary framework and need to be dealt with in a non-hierarchical, heterogeneous manner and with the engagement of several (academic and non-academic) actors throughout the process. Transdisciplinarity, thus, approaches relevant everyday problems, making the pursuit of scientific knowledge more socially responsible, on the one hand, and reaching deeper levels of transformation in science, on the other. This cycle ends in 1996 with the Manifesto of Transdisciplinarity by Basarab Nicolescu, which introduces the three fundamental axioms for the methodology of transdisciplinarity: the ontology axiom, the logic axiom and the complexity axiom, as a way to guarantee that it can advance and effectively contribute in the various spheres of its potential.

In 2010, Christian Pohl and Julie Klein, among others, assimilated and interpreted the various meanings of the new way of understanding knowledge, that is, as plural and contextualized, and they went further, with a well-defined social purpose in the pursuit of the common good. For them, transdisciplinarity is geared towards relevant social issues, the co-production of knowledge and, of course, the unity of knowledge. In this work, the long spectrum of meanings was approached as a *continuum*, which arises from the disciplinary issue and ends in the pursuit of the common good.

The contents of this chapter were aimed at presenting the pillars for the definition of transdisciplinarity. The ideas presented here are translated in the following words: transdisciplinarity is established as a concept which recognizes that the physical and cognitive structure required to understand and seek knowledge needs transformation in order to overcome the divisions that disregard other forms of knowledge, those that are beyond disciplinary boundaries. It seeks the unity of knowledge in the interacting parts immediately connected to the conditioned element and continues as required in order to shed light on the multidimensionality and multireferentiality of Reality, implicitly linked, in the pursuit of the common good. As a result, transdisciplinarity reinforces the relation between the whole and its parts (not only is the part in the whole, but the whole is in the part), maintains the duality within the unit (it uses the premise that there are two logics which are, at once, complementary and antagonistic; thus, they are crucial for the system to function) and distinguishes processes as being at the same time products and producers, in a recursive perspective, in which everything that is produced returns to what produces it in a cycle contrary to the idea of linearity.

In other words, transdisciplinarity is the quest for the unity of knowledge, beyond disciplinary boundaries, in order to capture the entire complexity of the multidimensional Reality and the multireferential of the conditioned element, based on the focus on relevant social issues, providing profound levels of transformation in higher education, with a view to the co-production of scientific knowledge aimed at the common good.

Chapter 3

TRANSDISCIPLINARY CO-PRODUCTION

3.1 Evolution of the Concept of Co-Production

3.1.1 Co-Production for Economic Development in International Relations by Maurice Byé and Henri Bourguinat

The concept of co-production is not new; since the beginning of the twentieth century, the term is used to refer to the industrial production of chemicals,[1] armaments[2] or films.[3] In the economic aspect, it probably began to be used with the work of Maurice Byé, in 1965, which points out co-production as an essential element of international relations, since international co-production means producing in common and, therefore,

> bring together projects, choices, means of action at the level of industry and markets [...] in addition to market equilibrium or imbalance compensation, a true 'federation of decision-making centers'. Thus, more directly than any other formula, 'co-production' can be an instrument of solidarity.[4] (Byé 1965, n.p)

1 This book suggests that perhaps the first use of the term in industrial chemical co-production occurred in the 1930s.

2 This book suggests that perhaps this was the first use of the term in the co-production of weapons: Cornell, A. H. International Codevelopment and Co-Production of Weapons: Some Conclusions and Future Prospects. *Naval War College Review*, 64–75, 1970.

3 This book suggests that perhaps this was the first use of the term in the co-production of films: IM, H. Significance of the movie co-production. *Chosŏnilbo*, 1957.

4 *mettre en commun des projets, des choix, des moyens d'action au niveau de l'industrie et à celui des débouchés [...]au-delà de l'équilibre de marché ou des compensations de déséquilibre, une véritable « fédération des centres de décision ». Ainsi, plus directement que toute autre formule, la «co-production » peut être l'instrument d'une solidarité.*

It was also in France that the term continued to be used in the economic sense of international relations for developing countries with the work of Henri Bourguinat (1968, 1969). The author argued that the concept of co-production is the bedrock of regional markets, and only when co-productions have multiplied, when irreversible links between partners have been built, can real integration be considered to exist.

3.1.2 Co-Production in Public Service Compiled by Jeffrey Brudney and Robert England

In the late 1970s until the mid–1990s, the term appeared in the United States associated with the Public Service, for example, in the works of Percy (1978), Whitaker (1980), Sharp (1980), Kiser and Percy (1980) and Brudney and England (1983), among others. Clearly, there are two perspectives in these two decades: in the first, 'the concept of co-production is based on the recognition that public services are the joint product of the activities of citizens and government officials' (Whitaker 1980, p. 242).

Brudney and England (1983) pointed to another perspective, that of economic relevance, particularly by Kiser and Percy (1980), who reopened the discussion about the distinction between 'regular producers' and 'consumer producers' introduced by Parks et al. (1981).[5] While the former group is responsible for producing goods and services for the purpose of exchange (usually money), the latter group is responsible for producing services to consume the resulting production. These conceptions are critical to checking whether combining them both is efficient from an economic point of view, as they depend on the extent of interdependence and substitution capacity in the production of a given good or service and on the relative wages of regular producers. Co-production, therefore,

> involves a mixing of the productive efforts of regular and consumer producers. This mixing may occur directly, involving coordinated efforts in the same production process, or indirectly through independent, yet related efforts of regular producers and consumer producers. (Parks et al. 1981, p. 1002)

5 This article is the result of an ongoing discussion between the authors during the years 1979 and 1980 and Parks took responsibility for putting the ideas together and publishing them. One the authors was Elinor Ostrom, who received the Nobel Prize in Economics in 2009 and whose ideas will be discussed particularly in the next item.

Brandsen and Honingh (2016) summarized that in the co-production of public services, services are not only provided by employees of public agencies, but also co-produced by citizens and communities. In this context, the authors seek, instead of a single integrating concept, the various possibilities of implementing co-production, considering that (i) for some services, co-production is not a matter of choice, (ii) even inherent, the participation of citizens will vary in the extent to which they are involved in a given project and (iii) there is no single phenomenon of co-production; in fact, there are core elements identified as: (a) co-production is a relationship between the employees of an organization and (groups of) individual citizens, (b) it requires direct and active contributions from these citizens to the work of the organization and (c) the professional is a paid employee of the organization, while the citizen receives remuneration below market value or no remuneration. With these elements, they offer a revised definition of co-production: ‘a relationship between a paid employee of an organisation and (groups of) individual citizens that requires a direct and active contribution from these citizens to the work of the organisation’ (p. 431).

3.1.3 Co-Production as the Basis for a Polycentric Political System of Local Governance Described by Elinor Ostrom

In this scenario that begins with economic development and advances to public services, the most representative conception of co-production is that of Elinor Ostrom, in 1996 – followed and expanded by, among others, John Alford (2009, 2014), Tony Bovaird (2007) and Victor Pestoff (2006, 2009) – whose application reached economics, political science, public administration, to name but a few fields (Brudney and England 1983). The classic definition of Ostrom (1996), featured in Crossing the Great Divide: Co-production, Synergy, and Development, says:

> By co-production, mean the process through which inputs used to produce a good or service are contributed by individuals who are not “in” the same organisation. The “regular” producer [...] is most frequently a government agency. Whether the regular producer is the only producer of these goods and services depends both on the nature of the good or service itself and on the incentives that encourage the active participation of others. If the regular producer is the sole producer of these goods and services, he depends All public goods and services are potentially produced by the regular producer and by those who are frequently referred to as the client [...] Co-production implies

that citizens can play an active role in producing public goods and services of consequence to them. (p. 1073)

The scenario for this conception of co-production dates back to the early 1970s, when Elinor and Vincent Ostrom founded the Workshop on Political Theory and Policy Analysis[6] in 1973 at Indiana University, with the aim of creating an academic environment for discussion on 'governance processes at the local, national, and global level [that] can be crafted to enhance human well-being, while promoting democratic principles and sustainable resource management'[7] (Indiana University Bloomington, n.p).

In 40 years of history, in the frequent meetings in the Workshop, experts have deeply investigated how governance arrangements affect the performance of government agencies on the basis of a view built on theoretical precepts and experience, which they believed to be 'the critical connection between ideas and what is accomplished'. She received Nobel Prize recognition 'for her analysis of economic governance, especially the commons', whose motivation was to challenge 'the conventional wisdom by demonstrating how local property can be successfully managed by local commons without any regulation by central authorities or privatization' (The Nobel Prize 2009, n.p.).

In addition to discovering and characterizing the principles of sustainable *commons*, after studying hundreds of cases in the historic book Governing the Commons (Ostrom 1990), Ostrom also reinforced the concept of co-production (Ostrom 1996) whose 'most important feature is the participation of both the provider (producer) and the user (consumer) of the good or service, while sharing responsibilities for public services'[8] (Pacheco et al. 2020, p. 103). She also said: *'No government can be efficient and equitable without considerable input from citizens'*. (Ostrom 1996, p. 1083)

Ostrom (1996) pointed out that an important characteristic of co-production is the set of conditions that are required so that it can occur. First, the technologies in use must generate complementary goods or products rather than merely a substitute, because this condition generates synergy, as co-productive inputs are legally owned by several entities. According to the legal framework, it must be adaptable in the interests of both parties, which can only happen with a polycentric political system rather than a monocentric

6 Workshop in Political Theory and Policy Analysis.

7 *Governance processes at the local, national, and global level can be crafted to enhance human well-being, while promoting democratic principles and sustainable resource management.*

8 Our translation.

(highly centralized) political system. A polycentric policy offers citizens opportunities to organize not just one, but many government agencies. Third, credibility must be one of the pillars of co-production through transparent and enforceable contracts between government agencies and citizens. And finally, the fourth condition concerns incentives, which encourage the engagement of employees and citizens.

The author cites a major observation by Putnam (1993), about co-production encouraging citizens to develop other horizontal social relations and social capital, leading to the understanding that co-production quickly results in different types of interaction. This perspective confirms Ostrom's view of synergistic results promoted far beyond what theory alone can provide, leading to the statement of Ostrom's Law put forward by University of Chicago's professor Lee Anne Fennell: 'a resource arrangement that works in practice can work in theory' (Fennell 2011, p. 9).

3.1.4 The Co-Production of Scientific Knowledge Proposed by Sheila Jasanoff

Growing from disciplinary roots, science and technology are rich in theory, methodology and societal applications. However, it does not suffice to produce for society; there is also a need to produce with society[9] and overcome S&T disciplinary divisions, encouraging a fruitful dialogue that takes into account the multiple realities contained in scientific knowledge and social practices. At the intersection of this union is the abundant concept of co-production, in which scientific knowledge is produced in a whole process that involves both the scientific method and the social context.

Jasanoff (2004c) described the objectives of co-production in terms of four structuring components: the vision of science in society and society in science (description); the explanation about how co-production mitigates the limitations of scientific progress (explanation); the analysis of emerging orders (normativity); and the prescription of action (forecast). Involved in this context, the author presents co-production as the 'constant intertwining of the cognitive, the material, the social and the normative' (p. 6) in various traditional scientific agendas. For example, the author points out that for political science, co-production offers new ways of thinking about power;

9 Here is a note on Citizen Science, that is, initiatives that include the participation of amateurs, volunteers and enthusiasts in large-scale scientific projects, usually limited to data collection. A new perspective begins to emerge with the Extreme Citizen Science ExCiteS, with the purpose of expanding the current scope of participation, allowing any community to develop research projects to deal with its own reality.

for sociologists, co-production represents new ways of conceptualizing social structures and categories that emphasize connection; for anthropologists, it means the cultural capacity to produce and validate knowledge and artefacts. Indeed, the author understands co-production much more as a language,[10] 'a way of interpreting and accounting for complex phenomena so as to avoid the strategic deletions and omissions of most other approaches in the social sciences' (p. 3).

Co-production is not just how people organize or express themselves, but also what they value and how they take on their responsibilities. Or in the words of Jasanoff (2004), co-production realizes that the 'ways in which we know and represent the world (both nature and society) are inseparable from the ways in which we choose to live in it' (p. 2). This involvement between science and society is what the author calls symmetry of science, which emphasizes the social dimension of scientific knowledge. It is also a criticism of the separation of facts, reason and methodology from culture, politics and subjectivity.

A decade earlier, Latour (1993) evoked the Principle of Symmetry to record differences, that is, co-production asymmetries; particularly, he was concerned that even though they may be similar in the principle of co-production, collectives may differ in size. By the way, collective is the term used to describe the association of (1) humans and non-humans and (2) society. Jasanoff (2004b) explained that it was Bruno Latour who introduced the term co-production in the scientific context, and claimed that the nature-culture division is a creation of human beings and, more specifically, of Western societies. His anthropology of science advocates the construction of systems that mix politics, science, technology and nature: '[y]et this is precisely the amalgam I am looking for: to retain the production of a nature and of a society that allow changes [...], but without neglecting the co-production of sciences and societies' (p. 134).

Jasanoff (2004b) also pointed out that a significant goal of co-production is to incorporate knowledge production into governance practices and to shed light on how governance practices influence the production and use of knowledge, which by the way does not happen by chance, but rather, by defined and well-documented trajectories. In fact, the quest for scientific knowledge based on co-production attests to the proposition of Mendelsohn (1977) that scientific knowledge is socially constructed, as advocated in the essay *The social construction of scientific knowledge*. In that work, the author questioned whether scientific knowledge can be

---

10 *Idiom of co-production.*

advanced without an analysis of the structural, social and cultural context of knowledge itself, and he stated:

> Science is an activity of human beings acting and interacting, thus a social activity. Its knowledge, its statements, its techniques have been created by human beings and developed, nurtured and shared among groups of human beings. Scientific knowledge is therefore fundamentally social knowledge. (Mendelsohn 1977, p. 4).

The author pointed out that science still develops, to a large extent, isolated from society, given the way it was instituted; it is unable to act in a multifaceted perspective. Also, a new approach, capable of including the broader social matrix, should be included in the cognitive processes. Still, Mendelsohn (1977) recognized that many efforts have been made to include social issues in scientific development; for example, Thomas Kuhn's (2012) work deals with the impact of major conceptual changes in science, based on the recognition of the social nature of the procedures through which knowledge is achieved. He also highlighted the unique contribution of Robert K. Merton (1938), who highlighted one aspect that shows the social and economic influences on science, which he called extrinsic influences on scientific research. It should be noted that the sociologist's reflection did not prosper in the following two decades and only resurfaced in the 1960s with Thomas Kuhn, followed by the thoughts of Bloor (1976), Latour and Woolgar (1979) and Collins (1985) (Jasanoff 2004a).

3.2 Transdisciplinarity as a Co-Production of Knowledge

3.2.1 The Sustainable Knowledge of Robert Frodeman

As pointed out by Klein (2013), transdisciplinarity introduces new forms of knowledge production. The linear model, based on disciplinary divisions, has limitations that make it difficult to overcome isolation on the one hand, facilitate integration on the other, and include the participation of other members of society. Not even the possible scenarios of interdisciplinarity favour the creation of structures required for the development of new knowledge, in view of their interdependence. Along these lines, Frodeman (2014) directly associated the concept of transdisciplinarity with the co-production of knowledge:

> transdisciplinarity, aka the co-production of knowledge, implies the recognition of limits to knowledge production. (p. 7) [...] to efforts to move beyond university walls and toward the co-production of knowledge between academic and non-academic actors. (p. 61)

The author pointed out that the change towards the co-production of knowledge is probably the most significant challenge facing the academy and, underpinned by this circumstance, he presented the concept of sustainable knowledge as a *core element* of the transdisciplinary co-production of knowledge.

The Brundtland Report,[11] produced by the United Nations World Commission on Environment and Development, defined sustainability in article 15 as 'a process of change in which the exploitation of resources, the direction of investments, the orientation of technological development; and institutional change are all in harmony' (Brundtland 1987, n.p). The same document also defined sustainable development as 'development that meets the needs of the present without compromising the ability of future generations to meet their own needs' (p. 41). And in this context, there is the Science of Sustainability[12] an interdisciplinary field, whose objectives are to

> basic understanding of the dynamics of human-environmental systems; to facilitate the design, implementation and evaluation of practical interventions that promote sustainability in specific locations and contexts; and to improve the links between relevant research and innovation communities, on the one hand, and relevant policy and management communities on the other. (Frodeman 2011, p. 109)

Frodeman (2011) drew on this concept to addresses the issue of sustainability of knowledge production in current scientific and technological development. The author argued that he is not advocating not delving into new knowledge; there will always be a need for research to meet the new possibilities that arise. The problem lies in the infinite production of knowledge, because most researchers (and people in general) do not need as much information as the volume that is currently being produced.

The dilemma faced by knowledge production made Frodeman (2011) point to the dangers of the 'bad infinity' described by Hegel in his nineteenth-century book Science of Logic (Hegel 1812). For Hegel, the bad thing about infinity is that it is endless; it has no conclusion , no goal. On the other hand, if it is infinite, there is a self-contained totality, which can be compared to

11 The idea of sustainable development, which has been discussed since the 1970s, was materialized in this United Nations' document, which was named after Gro Harlem Brundtland, president of the United Nations World Commission on Environment and Development in 1983 and 1987.

12 Sustainability Science Program at Harvard University's Center for International Development (CID)

the good part of finitude: something that has an end, a limit, that exists in one. In this perspective, for Frodeman, the university experiences a bad infinite, with the *laissez faire*[13] of unsustainable knowledge production, evidenced in the social results of research, which are not always evident.

The author poses the question of how knowledge institutions can be restructured in order to set clear objectives to knowledge production and how this would affect the status of science. He concluded with the certainty that the age of disciplinary knowledge may be ending, but the true form of sustainability is still unknown.

3.2.2 The Transdisciplinary Co-production of Merritt Polk

Klein (2013) also included transdisciplinarity in the approach that encourages participatory and collaborative research and the resulting co-production of knowledge. Researchers have had enough of pointing out big social problems, for example, the participation of non-academic actors in university research. Polk (2015) called transdisciplinary co-production the 'interactive and participatory modes of knowledge production [which] engage non-scientific actors and stakeholders by bridging disciplinary and sector-based boundaries' (p. 111). This integrative approach, which highlights the presence of actors outside the academic environment, shows that they are legitimate knowledge holders and that their experience is essential to knowledge creation.

The use of the term co-production, which can take on several specific meanings, is based on studies of science and technology in Jasanoff (2004c), to emphasize the joint responsibility of situated knowledge and scientific knowledge in the pursuit of problem solving, as an 'ontological condition that inundates all science-society interactions' (Polk 2015, p. 111). In this context, transdisciplinary co-production as derived from the articulation processes for knowledge creation is:

> *a research approach targeting real life problem solving.* Knowledge is co-produced through the combination of scientific perspectives with other types of relevant perspectives and experience from real world practice including policymaking, administration, business and community life. Co-production occurs through practitioners

13 Expression, derived from liberalism, which means 'let it be', used as a synonym for non-state intervention. The full version is '*laissez faire, laissez aller, laissez passer, le monde va de lui-même*', which is translated as 'let it be done, let it go, let it pass, the world will go by itself'.

> and researchers participating in the entire knowledge production process including joint problem formulation, knowledge generation, application in both scientific and real-world contexts, and mutual quality control of scientific rigor, social robustness and effectiveness. (p. 111).

It is worth highlighting the author's explanation for the term practitioner, which refers to individuals with public or private professional mandates, such as employees of public agencies and representatives of business or community groups.

Polk (2015) chose the transdisciplinary co-production structure of Mistra Urban Futures[14] to manage the challenges of applied research, which includes aspects of (i) inclusion and (ii) collaboration (to ensure mutual commitment and responsibility); (iii) integration (in order to identify and share different types of knowledge and experiences); (iv) usability (with the purpose of making explicit what can effectively contribute to problem solving, including the assessment of their social robustness) and (v) reflexivity (with a view to assessing the choices made on a permanent basis). Inclusion is specifically related to the right to knowledge resulting from the research of all stakeholders while collaboration particularly refers to processes, methods, intensity and level of participation.

Furthermore, he pointed out three challenges that are present across all phases of research. First, how to identify the priorities (and knowledge and practical needs) of a heterogeneous group and arrive at a common denominator of commitments and priorities, while maintaining enthusiasm throughout. The result of achieving shared responsibility with integrity is the quality and usability of research.

Second, just identifying the priorities indicated in the first challenge is not enough; the multiple demands of the actors need to be integrated. Polk (2015) suggested the social psychology approach, based on the meaning of organizational learning by Godemann (2008), in contrast to traditional approaches that focus on methods and tools for integration. At work, integration of knowledge is based on three main activities: (i) the exchange of information and knowledge; (ii) the creation of a common basis of understanding and (iii) the meta-reflexivity within groups, with the aim of

---

14 It is a research and knowledge centre that operated from 2010 to 2020, seeking to generate and enable the implementation of knowledge that promotes sustainable urbanization, through transdisciplinary research and knowledge co-production with local and global stakeholders. It was administered by a consortium of public institutions based in Gothenburg and hosted at Chalmers University of Technology.

revealing the perspectives of others through Game Theory, which has many implications for the study of cooperation between individuals. The three activities, within the co-production structure, lead to knowledge integration.

Third, how the results of co-production of knowledge can be incorporated in varied contexts. Here, Polk (2015) used the concept of social robustness, described by Nowotny et al. (2013), as 'the more strongly contextualized a scientific field or research domain is, the more socially robust is the knowledge it is likely to produce' (n.p).

3.3 Partial Considerations

The economic aspect of the concept of co-production probably started in the mid-1960s, with the work of Maurice Byé, who advocated the co-production of goods for improvement of international relations to facilitate regional integration, on the one hand, and as an instrument of solidarity, on the other. In the 1980s and 1990s, the concept was used in the United States in association with public service, particularly based on the recognition that these activities are the result of the combined action of citizens and government officials.

He collaborated with the American group, Elinor Ostrom, whose concern with the subject remained throughout her life, leading her to receive the Nobel Prize in Economics in 2009. In 1996, she created the most representative concept of co-production as the basis for a polycentric political system of local governance, in a process whereby the inputs used in the production of goods or services are provided by people who are not in the same organization.

In the 2000s, Sheila Jasanoff described the goals of co-production in terms of producing scientific knowledge with society, overcoming disciplinary divisions, encouraging dialogue between academic and non-academic actors, realizing the multiple realities contained in scientific knowledge and social practices. She went deeper into the issue and pointed out that a significant goal of co-production is to generate knowledge from governance practices and to shed light on how governance practices influence the production and use of knowledge. With the meaning given by the author, the question is regarded as transdisciplinarity of co-production or transdisciplinary co-production, with the research studies of Robert Frodeman (2014) and Merrit Polk (2015).

Frodeman's research reinforced the role of transdisciplinarity in new forms of scientific knowledge production and is geared towards the concept of sustainable knowledge, emphasizing the need for limits to its production, as a core element of sustainability and transdisciplinary co-production. Polk's research, in turn, inscribes transdisciplinarity in the approach that encourages participatory and collaborative research and the resulting co-production of knowledge, highlighting the presence of non-academic actors, recognizing that they are legitimate knowledge holders and that their experience is crucial to knowledge creation.

The content of this chapter was aimed at introducing the pillars for the definition of transdisciplinarity of co-production. The ideas presented are translated into the following words: transdisciplinarity of co-production is the vision of science in society and society in science, offering ways of interpreting and representing complex phenomena, in order to avoid strategic omissions by academic and non-academic actors for the resolution of relevant problems in society. As a result, it has become the most significant challenge facing the academy and the most necessary one, as it ensures the connection between disciplinary and sectoral boundaries.

In other words, the transdisciplinary co-production of scientific knowledge is produced in the convergence of the scientific perspective and the social perspective, overcoming the disciplinary divisions of science and technology, through the creation of a fertile field of dialogue, which perceives the multiple realities of science and social practices, involving both the scientific method and the context in which the problem lies.

Chapter 4

TRANSDISCIPLINARY RESEARCH

4.1 The Integrative Approach

Integration is central to all phases of transdisciplinary research, from problem identification and structuring to implementation. Pohl et al. (2018) argued that integration is not a value in itself, but a necessary methodology to meet society's demands for knowledge. Some authors, such as Bergmann et al. (2012), classified interdisciplinary and transdisciplinary research as integrative research, based on the degree of interaction between academic and non-academic participants. A specific feature of this classification is precisely the fact that it acknowledges the involvement of interested parties in integrative approaches, and this characteristic also poses the greatest challenge.

Disciplinary interactions, which evolve into interdisciplinary research, refer, in this *continuum*, to the degree of cooperation and integration of the various disciplines for resolution of the research problem. When the involvement of non-academic actors is added to this scenario, the *continuum* flows into transdisciplinary research (Tress et al. 2005). It is a trend that does not emphasize differentiation and specialization, but rather the need to unite various fields of knowledge and work experiences that can result in new specialties and types of socialization and, in the end, in more robust, resistant and healthier institutions.

Genuine transdisciplinary integrative research is concerned with coordinating different bodies of knowledge, after identifying gaps in the science, technology and society tripod, and understanding how to fill these gaps with appropriate scientific methodology and theoretical frameworks. It is the recognition that the bases of specialized knowledge are dispersed in the very heterogeneity of reality and that, therefore, only an integrative approach can possibly capture essential features of the context where the problem lies. It is about bringing to the scientific framework relevant social problems that need a solution and take science to the relevant problems that need a solution, wherever they are. It is about emphasizing the contextuality of transdisciplinary research, whose roots are founded on the needs of affected social actors, the inclusion of the participants involved and the commitment to mutual learning, and whose highest quality is 'to have a practical effect on the world beyond science' (p. 43).

Bergmann et al. (2012) underlined that there are three dimensions of integration that must be considered in transdisciplinary processes: (i) the communicative dimension, which consists of the different linguistic expressions and specific communicative practices of each participant, with the objective of developing a common discursive practice and ensuring mutual understanding and ease of communication; (ii) the socio-organizational dimension, responsible for relating the different interests and actions to guarantee integration and motivation for the project, and (iii) the cognitive-epistemic dimension, a linking agent between the knowledge bases (specialized and disciplinary), whose aim is to identify the state-of-the-art that surrounds the project and build typical integration methodologies and the knowledge required.

Vaegs et al. (2014) complemented the interpretation of these dimensions. As far as the communicative dimension is concerned, they added that the shared daily practice of research will generate a common terminology, integrated to that particular reality, which will promote ongoing transdisciplinary understanding, although the terminological differentiation remains essential to the respective parts that support the project. As regards the socio-organizational dimension, they reinforced the idea that transdisciplinary research is based on different stocks of (scientific and nonscientific) knowledge and different perspectives of results; therefore, the actual challenge to be faced is to combine approaches with a view to bringing them closer to all participants, making interests compatible with activities so as to build a collective identity. For the cognitive-epistemic dimension, they emphasize that by associating different types of knowledge and connecting scientific and practical knowledge, the different points of view in research must be mutually understood (which leads them, inexorably, to recognize their own limits), ensuring methodological integration and achievement of results of common interest. For Jahn (2008), there is a technical-factual dimension that creates an integrated normative system, responsible for bringing together the technical solutions of the project (and of the workflow) and making them viable in a sustainable and functional way.

Jahn (2008), Bergmann et al. (2012) and Vaegs et al. (2014) argued that these dimensions are essential to foster integration and to ensure that there will be effective transdisciplinary cooperation. Dimensions, however, are relatively general terms which must be instrumentalized by type of integration and integration methods, that is, by integrative procedures and instruments.

Furthermore, Hadorn et al. (2010) noted that a framework for integration must necessarily contain the following questions:

a) What and whom is integration for? What are the goals of integration? What is it intended for?
b) Integration of what? In which perspectives?
c) In what context is integration taking place? What is the focus of action?
d) Who will integration be implemented by? Will it be implemented by individuals, subgroups or a group?
e) How will integration be implemented? What methods will be used?
f) How will the result be evaluated? What are the measures of success?
g) Have the objectives been achieved? Have new perspectives been produced? Has it resulted in effective action? Have all the necessary elements been included in the integration process?

Importantly, a framework needs to address at least one of the following five integration strategies: dialogue-based, model-based, product-based, vision-based, and common metric-based. The Knowledge Acquisition Design Framework (KAD) will specifically use the dialogue-based integration strategy.

4.2 The Pursuit of the Common Good

The core of transdisciplinary research is the search for solutions to a socially relevant issue that leads to the common good. However, this greater objective of transdisciplinarity is not always explicitly stated in its definitions, as done by Pohl and Hadorn (2017) in their conceptual framework, in which they formulated the transdisciplinary project around four points: (i) understanding the complexity of the issue, (ii) taking the various perspectives on the subject into consideration, (iii) establishing the relationships of specific and generic (or practical and scientific) knowledge, as the case may be and (iv) developing knowledge and practices that lead to what is perceived as the common good.

The common good has been an important concern of philosophy, politics and law since ancient times. In politics, the common good has always been associated with higher purposes and a virtuous life as opposed to corruption and the pursuit of self-interest at the expense of the community. At an extreme level, the concept was associated with individual sacrifices in the name of the common good. In contemporary discourse, ideological connotations are present in the conception of the common good, which reinforced the debate about its validity.

According to the Stanford Encyclopedia of Philosophy (2018),

> 'common good' refers to those facilities – whether material, cultural or institutional – that the members of a community provide to all members in order to fulfill a relational obligation they all have to care for certain interests that they have in common [...] As a philosophical concept, the common good is best understood as part of an encompassing model for practical reasoning among the members of a political community. (n.p.)

That is, the members of a community maintain an urban relationship, because they know that part of what they will need for community life will be shared, so there are also close but not identical concepts of 'goods of public interest' or 'public goods'.

Despite the philosophical divergences of the concept of the common good, given its long history spanning more than two thousand years, it continues to be a foundation for the public and private dimensions of social life. Regarding public social life, it can be said that it 'consists of a shared effort among members to maintain certain facilities for the sake of common interests' while private life 'consists of each member's pursuit of a distinct set of personal projects' (Stanford Encyclopedia of Philosophy 2018, n.p.).

The most influential historical view of the common good is that of Aristotle, considered by many to be the father of the idea of a common good,[1] who made it the central concept of his book Politics[2] (Jaede 2017). In his work, Aristotle defined a political community, for example, a city, as a partnership:

> The citizens of a political community are partners, and as with any other partnership, they pursue a common good. In the case of the city it is the most authoritative or highest good. The most authoritative and highest good of all, for Aristotle, is the virtue and happiness of the citizens, and the purpose of the city is to make it possible for the citizens to achieve this virtue and happiness. (Clayton 1995, n.p.)

1 Plato addressed the theme before, in The Republic: 'although the specific term "common good" is not present in the texts of Plato, it is clear throughout his writing that he strongly believed a certain common goal existed in society and in politics. The best political order for Plato was that which promoted social peace in an environment of cooperation and friendship among different social groups, each benefiting from and adding to the common good)' (Simm 2011).

2 It is composed of eight books (I: 1252a–1260b, II: 1261a–1274b, III: 1275a–1288b, IV: 1289a–1301b, V: 1301b–1316b, VI: 1317a–1323a, VII: 1323b–1337a, VIII: 1337b–1342b).

In another volume of his work Politics,[3] Aristotle used the concept of the common good to distinguish six kinds of possibilities between correct constitutions, which are in the common interest, and incorrect constitutions, which are in the interest of rulers. 'The correct regimes are monarchy (rule by one man for the common good), aristocracy (rule by a few for the common good), and polity (rule by the many for the common good)' (Clayton 1995, n.p.). Nowadays, the common good is a fundamental concept of social morality and should be the reason for political action.

Jaede (2017) includes other ancient thinkers who also developed influential ideas about the common good, such as Cicero (106–43 BC): "a people[...] it is an assemblage of people in large numbers in agreement with respect to justice and a partnership for the common good" (n.p.). And also Thomas Aquinas (1225–1274), who offers the viewpoint of Christianity: '[a] tyrannical government is not just, because it is directed, not to the common good [*bonum commune*], but to the private good [*bonum privatum*].' (n.p.). The historical view also evokes the thinkers of the modern era, whose concepts of the common good have become more pragmatic, shifting from politics towards considerations more geared towards the material well-being of individuals, such as Thomas Hobbes (1588–1679), for whom 'the role of the state then became to ensure that individuals can pursue their personal ends in accordance with their common peace and safety, as opposed to promoting a moral vision of the good life' (n.p.), John Locke (1632–1704), for whom 'the government ought to respect and protect people's inalienable rights to life, liberty, and private property' (n.p.) and Immanuel Kant (1724–1804), who associated 'local realisations of the common good with a theory of international peace and cooperation' (n.p.).

Based on the social relationship between members of a community, five characteristics of the common good have been identified in many definitions by the Stanford Encyclopedia of Philosophy (2018). The first attribute is having a shared vision, focused on practical reasoning, in order to define a way of thinking and doing that reflects the mutual concern among members, through organized activities that can demonstrate the pattern in use. The second is having a set of common facilities that serve certain common interests, including institutions and social practices. It is actually worth mentioning that private property is part of this core, as a social institution, when it serves a common interest of citizens in being able to establish private control over their physical environment. The third is noting that the abstract interests of the privileged class must reach all citizens, for example,

3 Book III, Chapter 1279a-1.

> the interest in taking part in the most choice worthy way of life [...]; the interest in bodily security and property [...]; the interest in living a responsible and industrious private life [...] the interest in a fully adequate scheme of equal basic liberties [...] the interest in a fair opportunity to reach the more attractive positions in society [...]; and the interest in security and welfare. (n.p.)

The fourth aspect is having a solidary concern, a form of solidarity between those who are in the relationship, so that citizens can reason about the interests of others as if such interests were their own. A conception of the common good creates in its own reasoning what is in the interest of others, or in Rousseau's words, 'form of association that will defend and protect the person and goods of each associate with the full common force' (n.p.). And the fifth aspect is rejecting the aggregated view, that is, refusing to ignore the particular conditions of each member of the community, at the risk of corrupting the solidarity concern and imposing debilitating conditions on some of its fellow citizens in an integrated perspective.

The approaches to the common good can be summarized into three unitary theories, as named by Held (1970). First, the common good is objective and normative; it is not an object of discussion; it is something to be identified, but not invented. Second, there is no fundamental opposition between the common good and the individual good; both are contained in a common good, and if the individual good conflicts with the common good, the latter will prevail. Third, the common good is in the hands of good rulers, and the only threat is when they act in their own interests. (Simm 2011). Unitary theories lead to the reflection that '[t]he stronger the self-identification with the community, the less likely are conflicts between personal and communal interests' (p. 556).

Finally, the balance between public and private life raises important questions above all for politics (e.g. for the sustainability of democracy), for law (e.g. for social justice) and for philosophy (e.g. for the reflection of issues between public and private spheres), and among many others, when we should make decisions based on the common good.[4] Chomsky (2014) responded by pointing out that the unconditional concern for the common good is to encourage people to find ways to cultivate human development in its richest

4 Other questions that are part of philosophical discussions about the common good are why we should be concerned about the common good and what would be wrong with a community whose members withdraw from public life and focus exclusively on their private lives. Both issues are deeply rooted in the characteristics of today's information society.

diversity, questioning the social arrangements that lead them to the right, well-being and fulfilment of their fair aspirations.

4.3 The Conceptual Frameworks

Transdisciplinary research is organized on the basis of a conceptual framework that is essential for us to understand all aspects involved in research, the component parts and the way they interact. It plays a central role, above all in empirical and qualitative research, as it provides the necessary guidance to ensure scientific rigour and high-quality results. Another aspect worth highlighting is that the conceptual framework is a way of including the various levels of collaboration and participation. Collaborative research and participatory research are based on a variety of integrative perspectives in all of their facets. The framework, therefore, can be understood as a strategic mechanism to align the problem, objectives and expectations right from the beginning of a study, in processes that are based on the co-production of each step of the research process (Ravitch and Carl 2016).

For Ravitch and Carl (2016),

> The conceptual framework serves as a guide and ballast to research functioning as an integrating ecosystem that helps researchers intentionally bring all aspects of a study together through a process that explicates their connections, disjunctures, overlaps, tensions, and the contexts shaping a research setting and the study of phenomena in that setting. (p. 32)
>
> The conceptual framework helps you situate the study in its theoretical, conceptual, and practical contexts; includes implications related to the study's setting and participants; situates your researcher social location/identity and positionality; and articulates how all of these aspects are related to methodological frames and processes. (Ravitch and Carl 2016b, p. 62)

Preece et al. (2005) generically defined a framework as a set of concepts and elements that interrelate to describe how a system should behave, appear to be and be understood by users in the intended manner, while Crossan et al. (1999), in turn, explained that to be successful, the conceptual framework must present the territory that will be the setting of research and that must bring reality closer to theory based on (i) identification of the phenomenon to be studied; (ii) key premises related to the framework to be proposed; and (iii) the description of the interrelationships between the constituent elements of the proposed framework. Mcmeekin et al. (2020) argued that a conceptual framework includes a body of methods, rules of postulates, grounded principles, and a sequence of action.

There is a terminological confusion about the terms conceptual framework and theoretical framework and, in order to remove the ambiguity in this chapter, it should be clarified that both of them refer to the epistemological paradigm that a researcher adopts when looking at a particular research problem. The theoretical framework refers to the theory that researchers choose to guide them during their investigation. It is, therefore, the application of a theory, or a set of concepts extracted from one and the same theory, which allows the explanation of an event.

The conceptual framework, on the other hand, means that the research problem may or may not be contained in a single theory or in concepts within the same theory and, in fact, needs a summary of views, a series of related concepts, such as a way to link all elements of the research: 'researcher interests and goals, identity and positionality, context and setting (macro and micro), formal and informal theory, and methods (p. 26) [...] as a combination of experiential knowledge and prior theory and research (p. 28)' (Ravitch and Riggan 2017).

Ravitch and Riggan (2017) clearly showed that reality is always more complex than any theory can completely capture, and one needs to build a conceptual framework that takes this complexity into account, avoiding gross simplifications of the aspects involved. Indeed, the conceptual framework is not an aggregation of ideas, theories and methodologies. Much more than bringing a variety of specific knowledge, the parties must come together coherently, clarifying how they relate, based on a particular reference. Maxwell (2013) went further; he claimed that the conceptual framework organizes the structure of ideas and commitments that guide the study and, as an integration mechanism, the framework can be considered as an integrated set of elements, which works within and through 'the system of concepts, assumptions, expectations, beliefs, and theories that supports and informs your research' (p. 62).

Figure 4.1 shows the expected elements of a conceptual framework, in the view of Ravitch and Riggan (2017), as well as its main features.

Figure 4.2 expands and organizes the previous one, pointing out how the various elements of the research ecosystem intersect and influence each other.

The constituent elements of the conceptual structure presented in Figure 4.2 show that there are imperceptible bridges in the context, such as a 'connective tissue' (Ravitch and Carl 2016, p. 129), provided by the scope contained in the conceptual frameworks, which correlate methodologies with theoretical and situated knowledge, theory and practice, thus shedding light on various aspects. The framework helps to create an intentional process to connect the various parts of the context from a 'superstructure for the work [whose] [...] Personal interests and goals, identity and positionality, topical research, and theoretical frameworks each fit within that superstructure' (Ravitch and

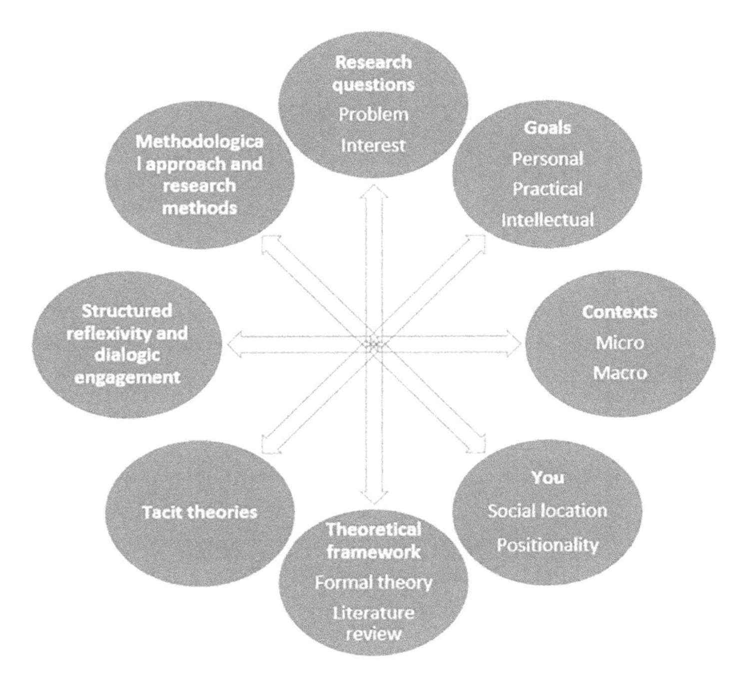

Figure 4.1 Essential elements of the conceptual framework.

Source: Adapted from Ravitch and Carl (2016b, p. 61).

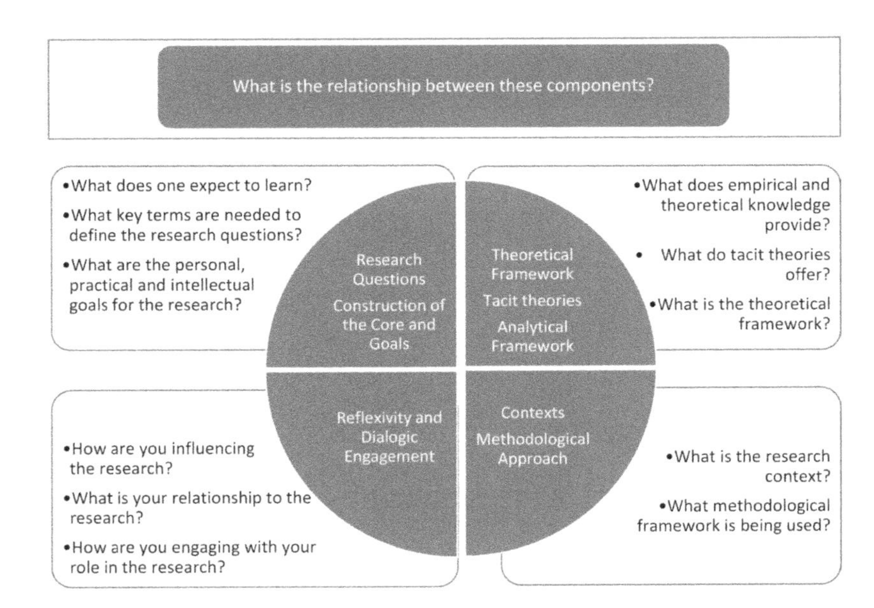

Figure 4.2 Essential interactive components of the conceptual framework.

Source: Adapted from Ravitch and Carl (2016a, p. 34).

Riggan 2017, p. 30). In the view of McMeekin et al. (2020), the benefits of using conceptual frameworks are: improved consistency of integration, robust objectives, accurate reports, high-quality research, standardized approaches and highly reliable findings.

Moreover, the conceptual framework provides the interconnection and interdependence between parts of a research study and even at the risk of oversimplification, these are the fundamental elements, which are not in order of priority and are not linear (Ravitch and Carl 2016): (i) the researcher, (ii) the theoretical framework – formal or established, (iii) tacit, situated and experiential knowledge, (iv) the research environment and immediate context, (v) the macro sociopolitical context, (vi) the problem situation, (vii) the research objectives, (vii) the methodology and (ix) working concepts. Some of these elements need to be clarified further.

In relation to the actual research environment and its immediate context (a specific location, an organization,[5] a group, a community or set of communities), it is worth noting that they influence the framework to a large extent, as they occupy the logical centre of the research. Some of the main questions that guide this characterization are:

- How environment and context influence goals.
- How the environment and context influence the methodology.
- What the interested parties are.
- Who is left out of the interested parties and why.
- How the environment and context define the theoretical framework.
- Transposition to a new environment and context is possible.
- How issues of power and hegemony are at play in the environment and context.
- What impact can issues of power and hegemony have on participant selection.
- What other aspects of environment and context seem important.

5 Although the KAD framework is related to the approach of Morgan's eight metaphors (1996) to understand an organization (as a machine, as an organism, as a brain, as a culture, as a political system, as psychic prisons, as flow and transformation and as instruments of domination), the literature review pointed to Jonathan Turner's definition (1997) as the most appropriate to the proposed conceptual framework: 'a complex of positions, rules, norms and values housed in certain types of social structures and organising relatively stable patterns of human activity with respect to fundamental problems in the production of life-sustaining resources, the reproduction of individuals and the support of viable social structures within a given environment' (p. 6). (Turner 1997; Morgan 1996).

Regarding the sociopolitical macro-context, the authors claimed that it is essential to understand the social, historical, national, international and global aspects surrounding research. They shape society and social interactions, influence the research topic, and affect organizational and personal structure and conditions. Indeed, there are two major implications of this element in research. The first is the importance of critically examining and understanding the sociopolitical nature of the issue and the way it reflects social and political aspects. The second is the importance of considering the historical moment that the research belongs to in order to orientate the vision and approach to the investigation. As noted by (Ravitch and Carl 2016), '[t]he process of continued contextualization comprises what we think of as a process of rugged contextualization, which refers to the rigorous pursuit of understanding context at these various levels and systematically appreciating how they inform and influence each other'[6] (p. 44). Some of the main questions that support this characterization are:

- What are the contexts (local, national, global) that shape the research?
- What does the local term represent?
- What are the segments (groups) at the local level?
- What criteria have shaped my perception of classification?
- What broader social and political contexts shape the research setting and in what ways?
- What policies are in force?
- What are the social constructions underlying the realities of the macro-context?
- How does the historical moment influence research?

Regarding the theoretical framework, it is worth clarifying its role as an integrative narrative between the theory or set of established theories in order to contextualize the focus of the study, the environment and context that shape exploration. They are formal because they are available in a variety of publication types that are recognized as accepted. Importantly, there is a need for a critical look at epistemological hegemony in order to ensure that all views are covered. Importantly, the theoretical framework is different from the literature review: the former is much more objective and precise in meaning construction, while the latter is the broader, early-stage process by which one can develop a specific theoretical framework.

6 *The process of continued contextualization comprises what we think of as a process of rugged contextualization, which refers to the rigorous pursuit of understanding context at these various levels and systematically appreciating how they inform and influence each other.*

The result is that throughout research and readings, the understanding of the theme is increased and refined, and ultimately challenges current theories. Some of the main questions that support this characterization are:

- What are the formal theories needed for research and why?
- How do they relate to research problems?
- In which field of knowledge is the research study inserted?
- What is the relationship between informal knowledge and formal knowledge arising from the theoretical framework?
- Can knowledge be treated in a polarized manner?
- Does the theoretical framework affect the environment?
- How does the theoretical framework affect the environment?
- What other disciplinary fields can complement the view of the problem?
- Does the theoretical framework reflect, in some way, the epistemological hegemony of the knowledge area?
- How could it enrich and broaden the dominant vision through the theoretical framework?

To conclude, Ravitch and Carl (2016) pointed out a series of functions of the framework, including: offer the general structure for reflection of objectives and target audience, provide guidance for an integrative methodological ecosystem, identify the relevance of selected theories and literature, provide the context of justification and disciplines involved, help refine and carefully design research questions, refine understanding of the researcher's position in context, help develop an appropriate framework for data collection, serve as a guide for selecting data analysis structures, be a basis for critically examining findings and analysis, help clarify how the theoretical framework intertwines with the various aspects of research, to be a mechanism for reflection on the importance and value of research and to consider the next research questions in context.

4.3.1 The ISOE model for the Transdisciplinary Research Process: Matthias Bergmann And Thomas Jahn (Germany, 2005)

The German model of the *Institute for Social-Ecological Research* (ISOE), proposed by Bergmann et al. (2012), presents three phases for transdisciplinary research. The objective of the first phase is to combine social and scientific problems, creating a common object of research. The second phase focuses on the integration and generation of new knowledge in order to use this diversity and develop viable solutions to problems. In the third phase, the results are integrated into the context and evaluated from the point of view of social progress and scientific progress, and as a result of the limits reached, new research questions will arise.

The model started in 2005 with the proposal of Jahn (2008),[7] which distinguishes three research approaches: everyday life, scientific and integrative modes. In the first mode, everyday life, '[T]he sciences accept, so to speak, an order from society (political sphere, economic sphere) to produce practical solutions to these problems' (p. 6). That is, a research problem will be analysed in a disciplinary (or multidisciplinary) way and its solution will be returned to society. Research aims to generate knowledge that is needed by those who demand it, and to identify new necessary knowledge, in a virtuous circle closed in on itself. Figure 4.3 summarizes the approach in everyday life mode.

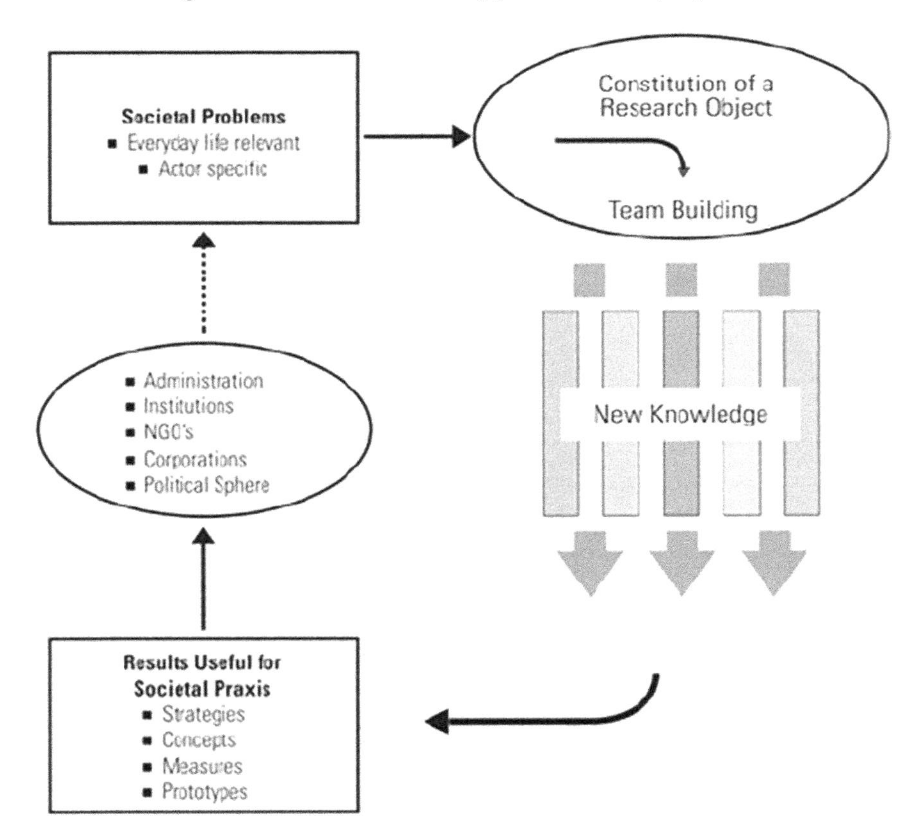

Figure 4.3 Research approach in everyday life mode.

Source: Jahn (2008, p. 6).

7 *The 2005 model was published in German, in Jahn (2005).*

In research processes in the scientific mode, the starting point is a complex problem internal to science, which involves theories, concepts and conceptions at the frontiers of knowledge. The aim is to increase scientific understanding and develop new knowledge. The virtuous circle closes again, as shown in Figure 4.4 of the scientific mode approach.

Often, however, the two approaches are inseparable and both paths must be followed simultaneously, giving rise to the integrative approach, which recognizes that societal problems are often problems of lack of knowledge about a particular issue. This approach, understood by the author as hybrid social problems, gives rise to integrative research, one in which both research paths are necessary. Here lies the specific characteristic of transdisciplinary research: the simultaneous search for two epistemic paths, guided by

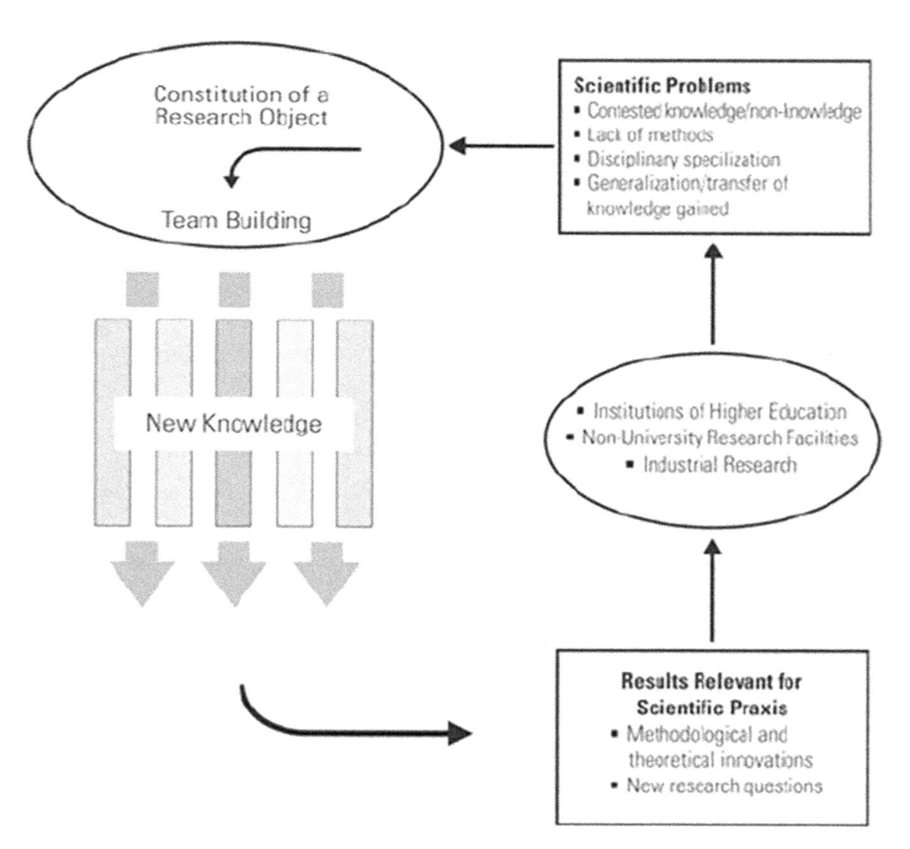

Figure 4.4 Research approach in scientific mode.

Source: Jahn (2008, p. 7).

the challenges of solving a problem with scientific stimulus (Bergmann et al. 2012). The authors call this result 'Transdisciplinary Added Value'.

> The transdisciplinary research process [...] is characterized by integration problems (epistemological, social, communicative and technological) and participative research arrangements (the inclusion of those affected, of users or stake holders, all in a process of mutual learning). (Jahn 2008, p. 9)

Bergmann et al. (2017) emphasized the need for a detailed model of the research project in order to (i) highlight upcoming activities throughout the project phases and (ii) ensure that the definition, analysis and implementation phases serve at least two different purposes: science and practice.

It starts with the problem of transdisciplinary research, which carries with it the idiosyncrasy of not being able to consider the question of the real world directly as a research task. The social problem needs a process of construction in scientific expression and understanding. The description of the problem, spontaneously placed (usually the starting point of the project) or academically structured, should be reformulated into a 'particular *epistemic object*, a scientific entity'[8] (p. 38). This step must be taken consciously at the beginning of a research study, as an integration strategy from the start, and must be understood as an agreement between the parties. In terms of integration, Bergmann et al. (2012) pointed out that terminological understandings begin here, new terms and concepts begin to be used and developed and are incorporated into common understanding. The project is already moving towards an integration methodology, whether existing or not, and the predominance of specific cultures becomes more difficult. In other words, 'work atmosphere and the scientific culture within the project [were] made more transparent and more easily acceptable to all parties involved' (p. 39). The authors call this approach *critical transdisciplinarity*,

> real-world problems and the corresponding scientific descriptions of these stand in a critical relationship to one another – each taking up a critical distance to the other. Scientific description distances itself from the societal problem, and vice versa; each testing the other. Transdisciplinarity is thus no affirmative approach – neither on the part of science, which must maintain a critical distance to the interest-determined descriptions of the societal problem; nor on the part of the societal actors involved, who must monitor the research process, critically viewing it with their expertise. (Bergmann et al. 2012, p. 39)

8 *Particular epistemic object, a scientific entity.*

The research planning is carried on with a clear framework in the *design* integration, which requires the absolute identification of the partners, whether academic and non-academic (or as Bergmann et al. (2012) refer to, scientific and practice partners). Integrative research strategies are diverse, but they all converge upon the need to create a concept of disciplinary integration, in order to guarantee the connectivity of the various knowledge bases in a collaborative process. In this scenario, one commonly works with interdisciplinary models as well as appropriate practical knowledge, establishing, in fact, a model of transdisciplinary research.

The last phase of the research process culminates in a provisional result, which brings together the new knowledge of practice and the expected scientific advance. In order to achieve the results, however, one should check the depth of the solution's effect, its relevance to the problem in question and the barriers that prevent the entire scientific and practical team from using it. In short, one should evaluate the results both in terms of the feasibility of the solutions proposed in practice and as a contribution to science, 'a conceptual clarification, which serves to make the complex results that have been developed in the course of the research process clearer, more understandable and better communicable' (Bergmann et al. 2012, p. 41). Figure 4.5 shows the everyday life and the scientific approaches.

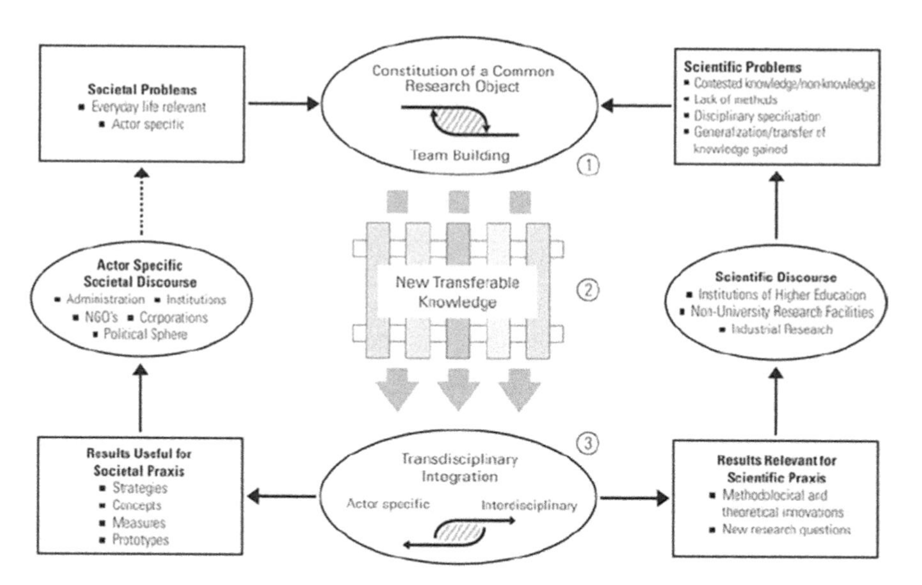

Figure 4.5 Research approach in integrative mode.

Source: Jahn (2008, p. 8).

4.3.2 Principles for designing Transdisciplinary Research: Christian Pohl and Gertrude Hirsch Hadorn (Switzerland, 2007)

Before proceeding, it should be noted that the common good refers to the underlying values in different thematic fields, which leads to specific contexts referring to the specific common good of a given community. Herein lies one of the great challenges of transdisciplinary research: identifying the concrete meaning of an underlying value in the research project and ensuring the compatibility of objectives between the various actors involved, in order to achieve harmony between credibility (scientific adequacy) and legitimacy (experience of interested parties) (Pohl et al. 2010).

The structuring elements of transdisciplinary research proposed by Pohl and Hadorn (2017), established in the 'Principles for Designing Transdisciplinary Research Framework',[9] seek to ensure that the issues are covered. Of the elements presented previously, the text only focused more deeply on the last topic, the common good, which was highlighted as the main objective of transdisciplinarity. The others need some additional comments.

Accepting the complexity of reality results in the imposition of collectively defining the problem. In transdisciplinary research, perceiving the problem from different perspectives means recognizing that what may be a problem for one person means nothing to another. The different interpretations require an attempt to achieve a consensus when determining the extent of the problem, through appropriate methodologies for this purpose, with 'methods that explore the different problem views by qualitative analysis and interrelate them by means of dialogue methods' (Pohl 2010, p. 70).

The next step is the integration of different parts, a central challenge of transdisciplinary research, considering that practices are 'founded in different value systems and different ideas about what relevant knowledge is, how it can be gained, or what role science should play in social change' (Elkana 1979 as cited in Pohl 2010, p. 70). Integration, in fact, is contained in the entire transdisciplinary research process, starting with the definition of the problem and extending to the incorporation of results.

The implementation of the research results 'should be seen as an intervention in a social system' (Pohl 2010, p. 70) but it does not end with the delivery of the solution to society, given that the 'effects should be carefully observed with particular attention to surprises (unexpected effects)' (p. 70) and to the impacts that can occur at various levels, whether intentional or not. In the words of Groß and Hoffmann-Riem, it is an 'experimental

9 Principles for Designing Transdisciplinary Research Framework.

implementation' (Groß and Hoffmann-Riem 2005 as cited in Pohl 2010, p. 70), which may require planning and conducting new interventions, in a long-term perspective, as well as recursive planning of the research process, going back and forth between implementing, analysing, developing new solutions and perhaps reframing the problem (Pohl and Hadorn 2017). In this condition lies the Principle of Reflexivity through recursion, and the authors stressed that

> *the three phases above are not a linear set of steps to be followed sequentially.* Instead, the complexity and interdependence of the phases means that what is learnt and decided in one phase affects the process of the other phases, so that the research process also involves revisiting earlier decisions and reshaping them in light of additional knowledge and insights. *Iteration allows targeted learning and helps avoid stalling in the face of complexity.*

In summary, this framework was proposed on the basis of the general principles and guidelines of the Network for Transdisciplinary Research (TD-NET) of the Swiss Academy of Arts and Sciences, and it aims to assist researchers working in the context of transdisciplinarity, throughout three phases of research: first, identification and structuring of the problem; second, problem analysis and third, fruition, that is, implementing the result of research into practice-oriented solutions for the common good. Figure 4.6 summarizes the three phases of the research by Pohl and Hadorn (2007).

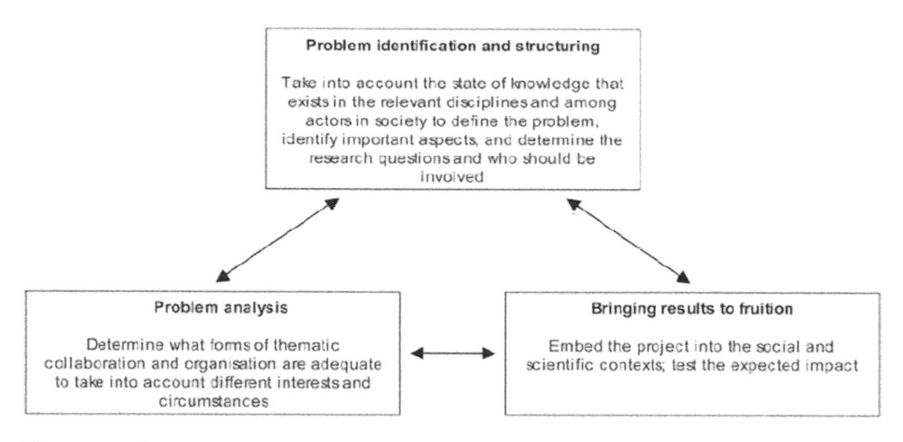

Figure 4.6 The three phases of transdisciplinary research.

Source: Pohl and Hadorn (2017, p. 230).

4.3.3 Analytical Structure for the Co-Production of Knowledge: Barbara Enengel, Andreas Muhar, Marianne Penker, Bernhard Freyer, Stephanie Drlik and Florian Ritter (Austria, 2012)

The Analytical Framework for Co-production of Knowledge by Enengel et al. (2012) was developed to provide further insights into the actors who contribute particular kinds of knowledge in each phase of a transdisciplinary doctoral research and what they contribute, considering the integration of scientific and non-academic knowledge to solve complex problems. Four main elements form the core of the structure.

The first (who) refers to who the actors are and what their functions are, as it does not suffice to know only who academics and non-academics are. The model prefers to classify non-academics into three levels: at the first level, practice specialists, usually employees of public agencies or non-governmental organizations. In general, they are familiar with the practical and political aspects of the issues investigated. At the second level, actors with formal or informal responsibilities, usually local politicians, local leaders or regional managers. At the third level, local actors, who are directly affected by the problem or who are involved in the case.

The second element (when) refers to research phases, since the co-production of knowledge has different intensities in the various phases of a project. Stakeholders are typically more involved in defining research questions and discussing results, sometimes in data collection but rarely in data analysis and subsequent publications. The model also points to insufficient information and even absence of the problem's history, which may indicate that there was no interaction at this stage.

The third element (why) is research objectives and rationale, since stakeholder involvement may vary. The different objectives of the interested parties must be clear in order to avoid disappointment throughout and at the end of the project; for example, the implementation of a solution may be outside the scope of the project, or it should be made clear that academics will be in charge of the methodology.

The fourth structuring element (what) is the type of knowledge present in the project and that will be generated as a result of it. The model organizes six types of knowledge, organized into three dimensions, which do not have a hierarchical or hegemonic relationship; on the contrary, they need to portray the possibility of assimilation, insertion, integration and union. They are:

- Scale. *Specific knowledge* of context, which refers to the concrete setting of the individual case. *General knowledge,* universally valid and systematically expressed, free from context-specific conditions and restrictions.

- Function. *Phenomenological knowledge,* which addresses local, social and environmental phenomena and their description. *Strategic knowledge,* which focuses on the connections and interrelationships of the elements of the system.
- Epistemology (Cognition). *Experiential knowledge,* derived from one's own life experience or traditional knowledge and often tacit or implicit and therefore generally not formalized or systematized. *Scientific knowledge,* based on empirical evidence or scientifically recognized theories; it is systematic, formalized and explicit.

The model by Enengel et al. (2012) is represented in Figure 4.7.

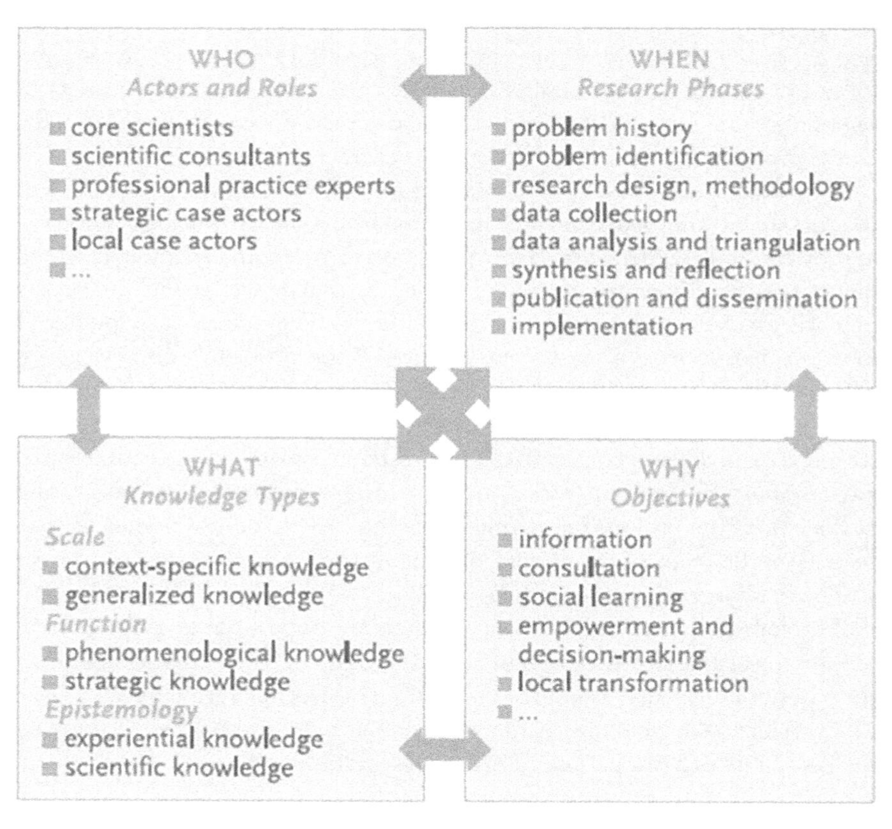

Figure 4.7 Analytical framework for co-production of knowledge.

Source: Enengel et al. (2012, p. 108).

4.3.4 Integrative applied research: Gabriele Bammer (Australia, 2013)

Among the challenges of academic research are the problems arising from the inherent complexity of reality and the need to solve problems that go beyond a discipline. Systems thinking and complex thinking are at the core of solving these questions through multidisciplinary, interdisciplinary and transdisciplinary research methodologies that maintain the benefits of discipline-based research while overcoming its limitations. Notably, there are other approaches with the same concern for integration, such as post-normal science, systemic intervention, integrated assessment, sustainable science, team science, mode 2 and action research among others. Bammer (2013) pointed out, however, that there is little cross-fertilization in these groups and teams tend to be isolated from one another.

Thus, she proposed a new approach that she called integrative applied research, which encompasses many of the methodological options above. It is characterized as a style of research (as with quantitative research, empirical research, experimental research and theoretical research) with the objective of solving complex problems in the real world and that can include several methodological options, recognizing that even in complex thinking, several approaches are possible and some are more suitable than others.[10] Bammer (2013) went further and actually proposed a new discipline, called Integration and Implementation Sciences (I2S),[11]' discipline that underpins integrative applied research and which develops and applies concepts and methods for knowledge synthesis, understanding and managing diverse unknowns and providing integrated research support for policy and practice change' (Bammer 2013, p. 18).

Integrative applied research has three domains: (i) the synthesis of disciplinary and stakeholder knowledge, (ii) the understanding and management of diverse unknowns, and (iii) the provision of integrated research support for policy and practice change. The first refers to the recognition that disciplinary knowledge alone is not sufficient; the knowledge of all stakeholders is very relevant: the stakeholders that are affected by the problem and those who are in a position to influence the problem. The combination of these

10 In this variety of possibilities, the author even recognizes the possibility of defining the research problem after the project has started: 'The advantage of starting from a common problem statement is that everyone is working to the same end, but the disadvantage is that achieving such agreement is time consuming and may reduce flexibility for considering new aspects of the problem that become evident as the research progresses' (Bammer 2013, p. 9).

11 Integration and Implementation Sciences (i2S).

types of knowledge, which will include, among others, epistemologies, values, interests and worldviews, will result in a comprehensive understanding of the problem. The second domain, managing the unknown, usually does not receive attention, but in integrative applied research, risks that exist owing to unintended consequences arising from uncertainties, inaccuracies, mistakes, distortions, incompleteness, deletions, taboos, etc. are taken into account when defining the problem that results from the unified view of the various stakeholders. The third domain, which deals with integrated research support for policy and practice change, shows that, in most projects, decision-making is based on a limited understanding of the issue, and therefore this domain represents assistance for teams to get to know how government, business and civil society operate and how research can contribute. The same holds true the other way around: the research results are input for the government, companies and civil society. To this end, Bammer (2013) recommends the following initiatives: make available what is known including successes and failures, inform about the unknown elements in the issue, express criticism of current and proposed policies and practices and recommend policy and practical initiatives.

Bammer's (2013) framework also provides five questions to stimulate systems thinking in project design, in each of the three domains:

a. What are the aims and who are the beneficiaries?
b. What is involved (systems-based approaches; scoping to identify all relevant disciplines and stakeholders, relevant unknowns and possibilities for action; boundary setting to establish the most important priorities from the options identified by the scoping process; framing decisions to define the problem and articulate courses for improvement; identification of value differences, especially value conflicts; effective harnessing of different expertise in the research team, in addition to management of differences that cause irritation and conflict)?
c. What methods will be used (dialogue, modelling, experiments, focus groups, building resilience, techniques to combat suppression and deception, ways to exploit the benefits of unknowns, co-production among others)?
d. What (political, economic, historical, cultural) circumstances might influence research? Is authorization necessary? What are the facilitators and barriers?
e. Have the best choices been made in addressing each of the four questions above?

Figure 4.8 represents Bammer's framework (2013).

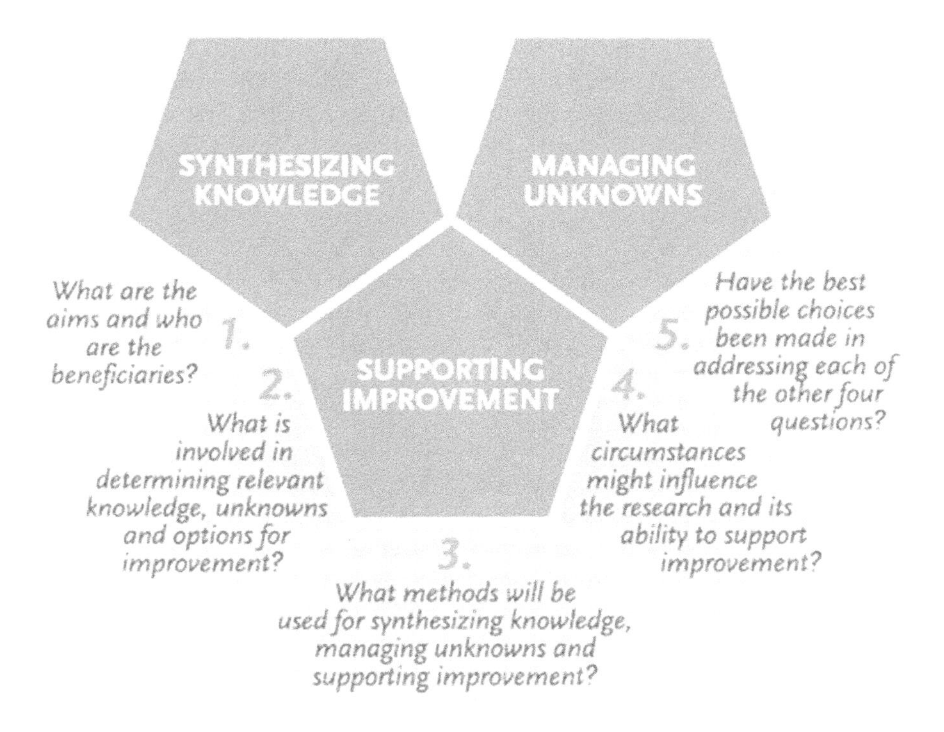

Figure 4.8 Identifying competencies in transdisciplinary research through I2S.

Source: Bammer (2019, p. 253).

4.3.5 Transdisciplinary Outcome Spaces: Cynthia Mitchell, Dena Fam and Dana Cordell (Australia, 2015)

The model of Mitchell et al. (2015) distinguishes three main domains of transdisciplinary outcome spaces for transdisciplinary research projects: situation, which is a field of inquiry; knowledge, which means the generation of relevant stocks and flows of knowledge, including academic knowledge and other forms of social knowledge; and transformational learning, which aims to increase the likelihood of change.

The model is rooted in the following attributes of transdisciplinary research: (i) *intention*: the intent of the research project; for example, is it aimed at improving the situation or understanding it?; (ii) *worldview*: the worldview or orientation of the research team, as this position influences the conceptual framework of the project; (iii) *experience and qualifications*: the qualifications, abilities, experiences and responsibility in the project, as these competences will effectively influence the project, especially in the quality of it; (iv) *funding arrangements*: who is funding the research, as this implies the commitments and expected results of it; (v) *degree of engagement across disciplines*: how is the theoretical framework understood and which current is adopted, as this results in the degree of transdisciplinarity needed; (vi) *past engagement with the situation*: the involvement of the team in the research situation, as this interferes with perceived trust and reputation, for instance; and (vii) *degree of engagement with the situation*: degree of stakeholder engagement, as it affects the breadth of perspectives included, including the legitimacy and credibility of research.

Based on the above scenarios, the model reaches the three transdisciplinary spaces of situation, knowledge and learning, all overlapping and interconnected. The situation refers to the research context, rejects the expression 'problem solving' and opts for 'problem situation', with the intention of showing that complex problems are rarely solved in a single project but, in fact, research continues. For the authors, situation is the best term because it definitely removes the expectation of closing the issue. Here, the following options are referred to for entering into the problem situation: Checkland's Soft Systems Methodology (SSM)[12] (Checkland 2000) based on Midgely's Critical Systems Thinking (CST)[13] (Midgely 2003).

In the situation approach, rather than making changes in the identified scenario, the situation space prefers to introduce improvements as a result of transdisciplinary research. A tangible improvement that can occur at different levels, for example, in strategy or in the introduction of a new work tool. It is hypothesized that "small changes, step by step, project by project, eventually lead to revolutionary changes" (Mitchell et al. 2015, p. 91). Other values mentioned in the model are the prolongation of the project effect that can go far beyond its end, the external factors that can provide windows of opportunity for improvement, and the project dynamics, which can alter its development. In short, for the transdisciplinary project to improve, the situation is the fundamental goal and, if possible, one should 'influence the broader situation towards a more sustainable trajectory [...] of these types of interventions, so that they can be repeated, refined and improved' (p. 92).

12 SSM are sets of linked activities that, together, can display the emergent property of the purpose. It is intrinsically linked to the concept of problem situation, which in turn leads to the perception of new knowledge and, consequently, to organisational learning (Checkland 2000).

13 It arose out of the need to analyze complex social problems, and established a variety of methodologies, such as SSM. It originated in systems thinking and that is why it inherited the concepts of system, element, relationship, boundary, input, transformation, output, environment, feedback, emergence, communication, control, identity and hierarchy.

The second space is about stocks and flows of knowledge, in a dynamic context of accessibility and inclusiveness; the latter is related to the intrinsic capacity of informational objects to be interpreted and used by the greatest number of people, and accessibility is related to the ability to disseminate knowledge. The flow in this model follows the thinking of Jantsch (1972); it 'relates to how knowledge moves: between disciplines; between theory and practice; between academic and professional practice; from within inside to outside of the project; up and down' (Mitchell et al. 2015, p. 92). The model of Mitchell et al. (2015) emphasizes that the production, sharing and dissemination of tangible knowledge artefacts are ways of extending the project's influence beyond its boundaries and that is, in fact, what is expected of transdisciplinary research: the production of knowledge that considers complexity and contextualization of social issues.

The last domain of the transdisciplinary space is that of mutual and transformational learning. Everyone involved in the project, at the end of their participation, must come out of the experience 'with new perspectives, new orientations, new strategies, and new tools – seeing and doing things differently as a result of their experience of transdisciplinary research' (Mitchell et al. 2015, p. 92). Learning as derived from the project, generated in a participatory and collaborative way, occurred as a result of dialogue and social interaction between team members, which led to reflections that only occurred because of the previously set up environment. It reached each team member individually and mutually; in this sense, the model is aligned with the concept of *transformative learning* proposed by Taylor (1998).

> Transformative, higher order, 'conceptual', 'generative' learning involves changes in norms and values, redefining goals that govern the decision-making process, reviewing and adjusting problem definitions (or perceptions of real-world situations), strategies, and actions of organisations and individuals involved. Transformational learning as defined in this framework denotes learning that leaves a legacy and contributes to changing the situation. (Mitchell et al. 2015, p. 93)

The conceptual map of outcome spaces highlights the context of the transdisciplinary project in a given field of vision, which exists as a result of the research team's experiences, knowledge and worldviews. Concomitant to the field of vision are uncertainties originating in the unknown. In addition to these, the project is also limited in time and space, requiring results compatible with available resources and with realistic expectations. Figure 4.9 shows the elements presented previously.

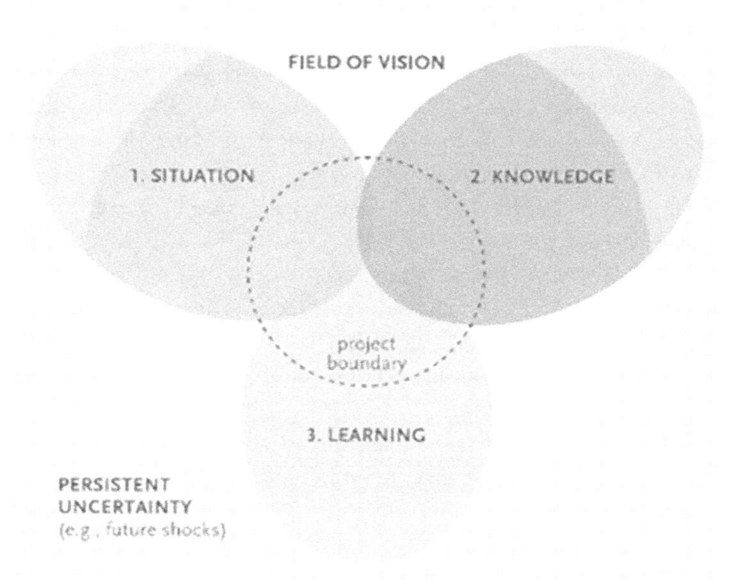

Figure 4.9 Conceptual map of outcome spaces.

Source: Mitchell et al. (2015, p. 91).

4.3.6 Four building blocks of systems thinking: Derek Cabrera and Laura Cabrera (United States, 2015)

The *Systems Thinking Made Simple* model proposed by Derek Cabrera and Laura Cabrera (2015) is based on the foundations of systems thinking, specifically on four principles that the authors deemed to be universal for understanding and practicing systems thinking: making *Distinctions*, organizing ideas into parts and sets of *Systems*, identifying *Relationships* and adopting *Perspectives* (DSRP). Cabrera and Cabrera (2015) argued that the DSRP offers a unifying and organizing principle for systems thinking and an indispensable analytical tool to solve complex problems.

These principles, in turn, constitute four simple rules of systems thinking. The Distinction Rule expresses that any idea or thing can be distinguished from other ideas or things, the Systems Rule expresses that any idea or thing can be divided into parts or grouped into a whole, the Relationship Rule expresses that any idea or thing can relate to other things and the Perspective Rule expresses that any idea or thing can be the point or the view of a perspective.

The model used the four rules to create four building blocks. The first block is called Distinctions (Identity-Order) and refers to the striking characteristic of systems thinking, which is to make distinctions (identity-other) between things and ideas (which will be called 'elements' from here on in this book).

How the boundaries of a system are defined is essential for understanding it, because the boundaries, while defining an element, define what the element is not, which at the same time leads to the definition of the other. Cabrera and Cabrera (2015) argued that systems thinkers consciously use these boundaries to challenge existing norms, labels and definitions and to identify biases in the way information is structured.

The second building block, Systems (Part-Whole), refers to how systems thinkers organize elements into part-whole systems to create meaning, since changing the way the elements are organized changes the very meaning of the elements. The act of thinking is definitely based on the grouping and separation of elements, according to some tacit or explicit criterion. Those who think in a systemic way naturally need to permanently analyse the larger context of which the elements are a part of.

The third building block is that of Relationships (Action–Reaction) between the elements. Systems thinking does not occur without an understanding of how the parts and the whole are related, thus creating a dynamic interaction between the elements – the most widespread of which is feedback to understand the reciprocal relationships.

The fourth block refers to Perspectives (Point of View), that is, the possibility of observing the element from various perspectives, from the particular perspective adopted in the choice of distinction, organization and relationship. Systems thinkers use the possibility of perspective to rethink distinctions, organizations, and relationships, and ultimately conceptual perspectives of choice. Figure 4.10 shows the principles and the respective building blocks.

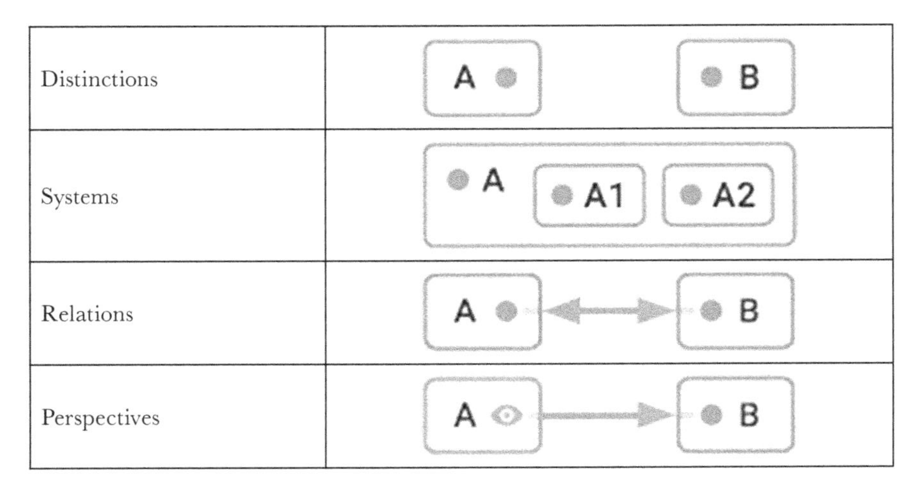

Figure 4.10 The four building blocks.

Source: Cabrera and Cabrera (2018, p. 200).

Once the building blocks are established, the model goes on to combine them to create thought patterns, as shown in Figure 4.11.

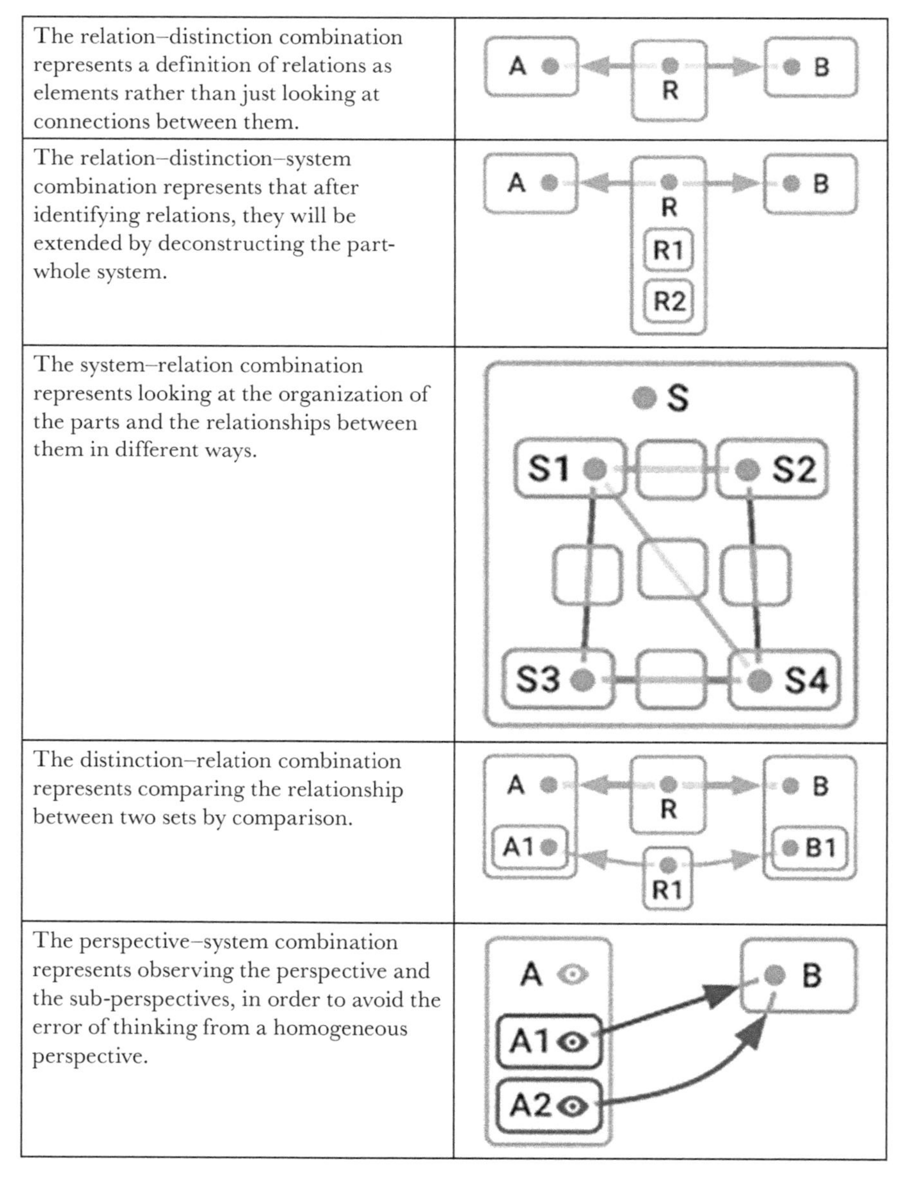

The relation–distinction combination represents a definition of relations as elements rather than just looking at connections between them.	
The relation–distinction–system combination represents that after identifying relations, they will be extended by deconstructing the part-whole system.	
The system–relation combination represents looking at the organization of the parts and the relationships between them in different ways.	
The distinction–relation combination represents comparing the relationship between two sets by comparison.	
The perspective–system combination represents observing the perspective and the sub-perspectives, in order to avoid the error of thinking from a homogeneous perspective.	

Figure 4.11 Combinations between building blocks.

Source: Cabrera and Cabrera (2018, p. 200).

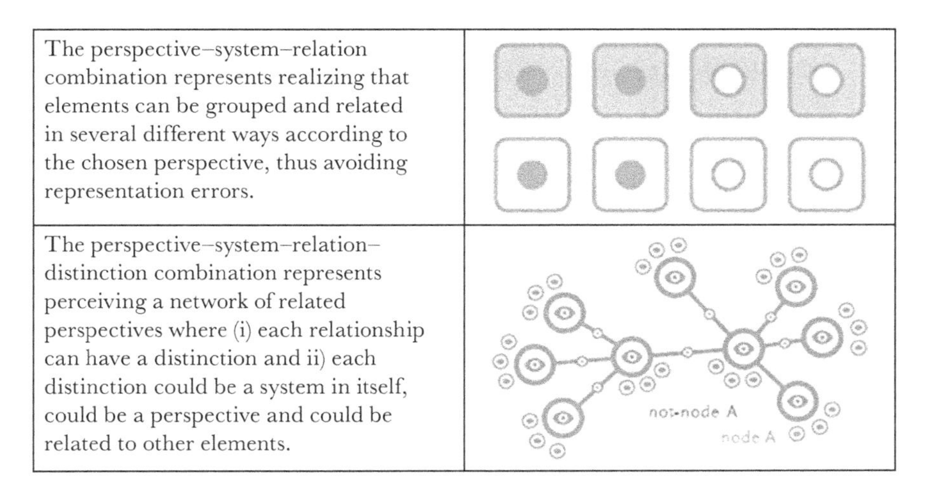

The perspective–system–relation combination represents realizing that elements can be grouped and related in several different ways according to the chosen perspective, thus avoiding representation errors.	
The perspective–system–relation–distinction combination represents perceiving a network of related perspectives where (i) each relationship can have a distinction and ii) each distinction could be a system in itself, could be a perspective and could be related to other elements.	not-node A node A

Figure 4.11 (*Continued*)

Effectively, this alternative became more widely available owing to the powerful modelling and visualization tools, given the complexity of the mathematical formulations behind the combinations. For explanatory purposes, the equation that arises from the condition that complexity is the collective action of the DSRP (Figure 4.12), derived from simple rules and interactions, is[14]:

$$ST_n = \bigoplus_{\text{agents } j \leq n} \otimes \{: D_o^i \mathrm{O} S_w^p \mathrm{O} R_r^a \mathrm{O} P_v^\rho :\}_j$$

Figure 4.12 Proposal for a systems thinking equation derived from the DSRP action.

Source: Cabrera et al. (2015, p. 538).

14 The equation explains that autonomous agents (information, ideas or things) following simple rules (D, S, R and P) with their elemental pairs (i-o, p-w, a-r and ρ-v) in nonlinear order (:) and with various co-implications of the rules (ᴑ), the collective dynamics of which over a time series j to n lead to the emergence of what we might refer to as systems thinking (ST) (Cabrera et al. 2015, p. 538). (*The equation explains that autonomous agents (information, ideas or things) following simple rules (D, S, R and P) with their elemental pairs (i-o, p-w, a-r and ρ-v) in nonlinear order (:) and with various co-implications of the rules (ᴑ), the collective dynamics of which over a time series j to n lead to the emergence of what we might refer to as systems thinking (ST)* (Cabrera et al. 2015, p. 538).

4.3.7 Context of the Interaction between Research and Government Policy: Vanesa Weyrauch (Ireland, 2016)

The *Knowledge Into Policy* model created in 2016 by the group of Weyrauch et al. (2016) proposes a transdisciplinary research framework that identifies strategic areas for the process of change, recognizing that policy is an essential element when aiming for transformation. The approach comprises six important dimensions to improve the use of knowledge in public policies: (i) macro-context; (ii) intra and inter-relationships with state and non-state agents; (iii) culture; (iv) organizational capacity; (v) management and processes and (vi) core resources. The framework allows users to systematically and comprehensively assess where the potential for change in government policy is the greatest, as well as where the most significant barriers to solving a research problem are.

On the first, there is significant literature on how the macro-context affects efforts to promote better use of knowledge in policies, usually focused on large contextual factors that go far beyond this perspective. This context refers to the structural and circumstantial factors at the national level that set the scene in which policies are made and include political, economic, social and cultural systems. These forces, in turn, can be classified into (i) structural factors (power distribution, participation and responsibility, knowledge regime, planning culture, discretional decision making and corruption, political competition, government effectiveness, priorities and transparency, intellectual environment and civil society, freedom of the press, among others), which rarely change significantly and can be considered as the most constant and regular external configuration of policies and (ii) circumstantial factors, which from time to time can emerge as a driving force of changes, such as popular pressure, crises and transitions.

The definition of Knowledge Regime is particularly important at this point; in the words of Campbell and Pedersen (2014), it is 'the organisational and institutional machinery that generates data, research, policy recommendations and other ideas that influence public debate and policymaking' (p. 6). Based on this understanding, there are several aspects to consider, that is, the information produced by a knowledge regime for policymakers involve ambiguities and uncertainties, which leads to the need for meaning construction, a process that, in turn, involves 'varying degrees of competition, negotiation and compromise – often involving power struggles – over the interpretation of problems and solutions for them' (p. 6). According to Garcé (2015), the knowledge regime includes the social valuation of science, which in turn also has a great influence on the supply and demand for research.

> In countries where the predominant culture is more rationalist, the demand for research tends to be greater and there is general acceptance that science can be neutral. On the other hand, in countries with a political culture that is suspicious of expert knowledge, there will be less demand for scientific knowledge, and science will tend to be used instrumentally as a means to an end. (p. 21)

Intra- and inter-institutional relationships are part of the macro-context, but they deserve special attention. There are two specific types of relationships that have a significant influence on how knowledge interacts (or not) with policies. The first is the internal relationships between the government institution and other related government agencies. The second concerns the interaction between the government institution and knowledge users and producers that can affect or be affected by the design and implementation of policies.

As regards relationships, the flow of information at different levels is noteworthy, as it may be limited by the degree of political affinity or the gap between political parties; as a result, there may be more or less information sharing. Likewise, the level of trust influences the depth of interaction and, by observing this condition, one can tell whether relationships are hierarchical or horizontal, rigid or flexible, or whether the relevant knowledge produced will be affected. In fact, the authors stated that 'institutional silos can limit access to research and evidence use' (Weyrauch et al. 2016, p. 36).

In addition to these situations, there is a lack of coordination between the government institutions involved, which can significantly impede the sharing of knowledge generated by research. On the other hand, it is worth noting that as coordination is achieved, the co-production of knowledge is fostered between different governmental and non-governmental actors, including academic ones. In this circumstance, windows of opportunity emerge for transdisciplinary research.

The other four remaining dimensions are key aspects that explain how a government institution thinks and acts, and they are presented in terms of how they interfere in the relationship between research and policymaking. It starts with culture, a set of values and assumptions accepted by a group and, therefore, passed on to new members as the correct way to perceive, think and feel the organization, including its problems. The authors used Schein's definition of culture (2004, p. 17 as cited in Weyrauch et al., 2016, p. 40).

> A pattern of shared basic assumptions that was learned by a group as it solved its problems of external adaptation and internal integration, that has worked well enough to be considered valid and therefore taught to new members as the correct way to perceive, think, and feel in relation to those problems.

This culture creates the daily context for practice and, in fact, the authors consider that there are several different cultures in the same organization, often conflicting, implicitly reflecting different values or worldviews. This condition leads to undesirable working relationships and clearly affects how research will be conducted and what knowledge will be used. Culture, therefore, needs to be well understood to determine the potential for political change in a government organization as a result of the knowledge gained from research.

In turn, organizational capacity concerns the ability of an organization to use its resources to execute its objectives and fulfil its mission; in the case of a public organization, it is the capacity to plan and implement public policies. In this understanding, there are two main components: human resources and the legal framework. In relation to human resources, the authors distinguish civil servants, more likely to preserve existing policies, and appointed individuals, prone to innovation, a fact which generally creates the need to realize that the former should be engaged in the changes arising from research studies as much or more than latter; that new knowledge must be incorporated into the *modus operandi* of the organization, thus promoting the design of communication and decision-making strategies based on the use of research findings and results. The issue of human resources leads to the need to understand the importance of leadership in promoting knowledge-based change in policies. The authors claimed that without commitment and the interest of leaders, it is impossible to change. As far as the legal framework is concerned, Weyrauch et al. (2016) clearly showed that the legal basis for change should not be underestimated, as all stakeholders, including research, are limited by rules and regulations. They warn that these sometimes complex milestones can immobilize research studies and smaller institutions.

Organizational management and processes refer to how each government institution organizes its work to achieve its mission and objectives, from planning to evaluation; how actions are implemented to enable members to fulfil their roles and responsibilities and how policy discussions and decision-making take place. Lusthaus et al. (1995, as quoted in Weyrauch et al. 2016) defined internal management systems as the mechanisms that guide interactions between people to ensure that the work in progress gets done. Facilitators to use knowledge as the foundation of 'planning, communication, decision making, problem-solving, monitoring, and evaluation' (p. 52) include consistent policies, and regulatory and budgetary frameworks that support R&D institutions.

Finally, core resources include budget, time, infrastructure and technology, broken down into other critical dimensions. Among the core resources, one naturally should emphasize the need for a good knowledge infrastructure in government aimed at public policymakers, with knowledge management strategies that maximize the use of informational resources on previous policies and contextualize the information in institutions and in research. In fact, the entire spectrum of tacit and implicit information that is present in public institutions must be organized: information contained in processes, relationships, products, people, communities of practice, epistemic communities – in short, all organizational intelligence must be identified and explained, based on reliability and availability. If knowledge is not on solid ground to advance, it will not be successful and will not contribute to policy making.

Budget constraints, in turn, are critical for the generation and use of knowledge from research. This limitation leads to inappropriate methodologies, civil servants unprepared for reflection and analysis, inability to hire researchers or suitable research centres to carry out specific studies, which is aggravated by the dispute for resources between research and the political agenda. It is noteworthy that financial containment not only impacts the generation of knowledge, but also the implementation of discoveries and results. Technological infrastructure, in turn, plays a fundamental role in the production and flow of knowledge relevant to public policies, with emphasis on the migration of information (already available in other media) to digital platforms. It should be noted that just having technologies is not enough; one needs to know how to use and effectively make use of the available knowledge.

Lack of time also inhibits policymakers' ability to use knowledge as a basis for decision-making. In practice, most of the deadlines for taking decisions are not compatible with the availability of the corresponding research results. Therefore, we need to ensure information flows that can meet decision-making needs, and to distinguish the decision-making typology of organizations, which must include decisions taken on a regular basis and those taken on specific occasions. Figure 4.13 shows the interrelationship between dimensions.

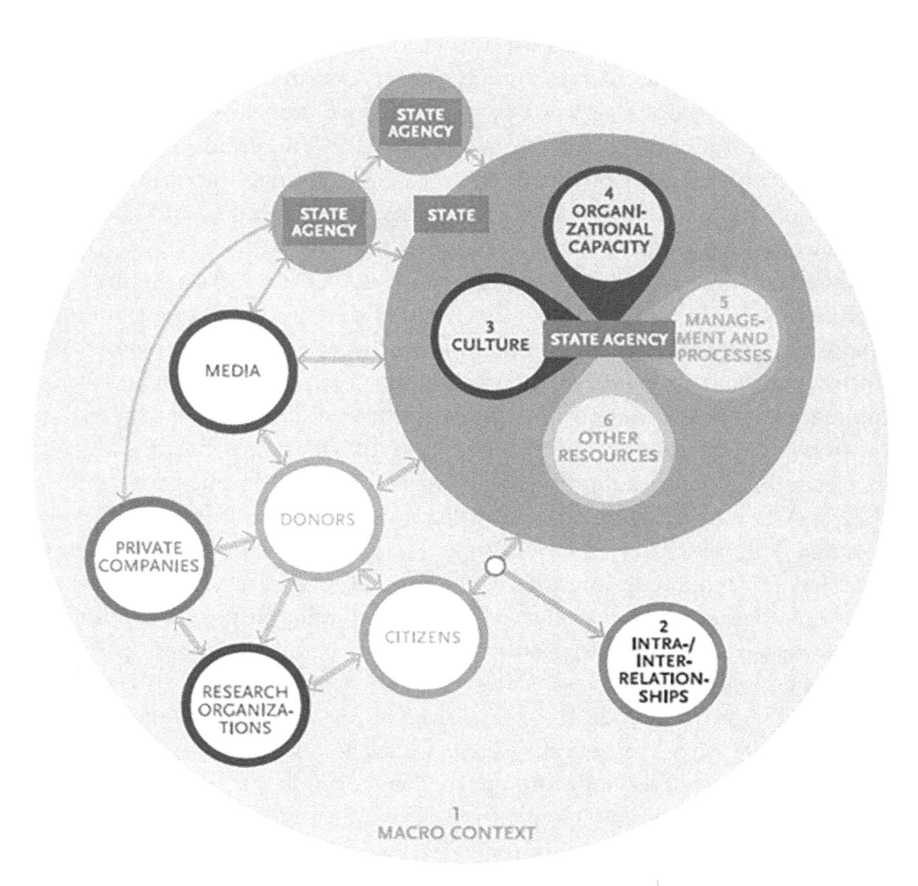

Figure 4.13 Main dimensions of the interaction between knowledge and policy.

Source: Weyrauch et al. (2016, p. 24).

4.3.8 Ten fundamentals to Contribute to Transformational Change: Ioan Fazey (Worldwide, 2018)

There are many types of research that are at the interface of scientific knowledge and knowledge arising from practice, particularly oriented towards social impact, co-creation of results, promotion of learning and generation of new knowledge closely interconnected between academics and non-academics. The authors call this typology second-order transformation research (Fazey et al. 2018), which makes explicit the assumptions and underlying characteristics of many research approaches, such as transdisciplinarity, mode 2 or action research.

This conceptual framework presents ten guiding foundations for solving complex and urgent problems through transdisciplinary second-order transformation research, based on the premise that researchers do not remain independent of the context; on the contrary,

> Through theories, concepts, and findings researchers also influence society [...] researchers are inevitably embedded within, and not separate from, the systems they seek to observe [...] also, arguably always interveners [...] the action is influenced by knowledge, including perceptions, implicit understandings, conscious and unconscious motivations, as well as values, morals, ethics and norms and behavioral habits. (Fazey et al. 2018, p. 56)

By the way, this is the foundational concept of second-order science, which arose with the idea of second-order cybernetics, when the observer is included in what is observed, whose consequence is to see science as an active part of social systems, becoming a concept of co-evolution of science and society (Umpleby 2014).

The first three fundamentals are related to research planning and objectives, while the others are related to research practice.

(i) Focus on transformations: research needs to focus on transformative changes rather than efforts on incremental or peripheral changes. For this to occur, governance, power, values, cultures and technology must receive equal attention.
(ii) Focus on solution processes: understanding the problem and defining the objectives is only one way to achieve the results; therefore, value can be placed on the results and on how they are expected to be achieved.
(iii) Focus on 'how to' practical knowledge: above all, the transformation approach requires a commitment to change, which calls for the involvement of academic knowledge (episteme) and practical knowledge; the latter is subdivided by the authors into technical knowledge ('know how' knowledge, techne) and practical knowledge (about knowing how to do it, phronesis), based on the definition of the ancient Greek philosophers.[15] For the authors of the model,

15 Very briefly, philosophers of antiquity defined *episteme* as scientific knowledge, *techne* as technical knowledge, *phronesis* as practical knowledge, *doxa* as an opinion, *sofia* as wisdom and *nous* as intelligence or rational thinking.

one needs to understand three types of practical knowledge: first, research into practice, in which researchers observe what the object of study is like (techne); second, research as practice – researchers (or practitioners) develop a process of change or new technologies through experimentation and iteration (phronesis); and third, research through practice, in the sense of developing the practice (techne and phronesis). At this point, the authors pointed out that '[r]esearch through practice is generally lacking in the humanities, social sciences and sciences, although it may sometimes emerge in transdisciplinary and action research' (Fazey et al. 2018, p. 62).

(iv) Approach the research from within: research needs to be approached in the context of the system in which it is inserted in order to make it easier to recognize the elements that contribute to the problem. From this perspective, research encourages interconnected approaches to action and learning, increasing the likelihood of innovation, learning and transformation.

(v) Work with normative aspects: research needs to find ways to work in the complex and multiple reality, incorporating values, ethics and aesthetics as part of the knowledge production process.

(vi) Seek to transcend current thinking and approaches: many contemporary problems cannot be addressed by the same kinds of thinking that created them. New ways of thinking are needed to make room for new questions, insights and solutions that can transcend current paradigms and disciplines.

(vii) Take a multifaceted approach to change: different paradigms, methodologies and methods affect the interpretation of phenomena and the way subsequent actions are prescribed, so multiple perspectives must be considered.

(viii) Acknowledge the value of alternative roles of researchers: researchers can be facilitators, mediators or catalysts in various segments of the project, which makes their actions more flexible and enhances the integration between stakeholders.

(ix) Encourage experimentation: transdisciplinary transformation research requires experimenting with change processes through projects and other local and context-specific initiatives.

(x) Be reflexive: reflexivity is the critical exploration of how circumstances influence interpretations, approaches and learning, whether they are 'perceptual, cognitive, theoretical, linguistic, political and cultural' (Fazey et al. 2018, p. 66). This includes widespread scepticism about a person's knowledge and value stances. One should also consider the opinions of others, and make the underlying values and assumptions explicit.

In summary, all fundamentals are important, but significant impact to research results is enhanced when they are applied together, as they 'will create a highly adaptive, reflexive, relational, collaborative and impact-oriented' form of research (Fazey et al. 2018, p. 66), as well as an 'intellectual depth in ways that expand the explicit and normative aspects of research' (p. 66). Figure 4.14 shows the ten principles.

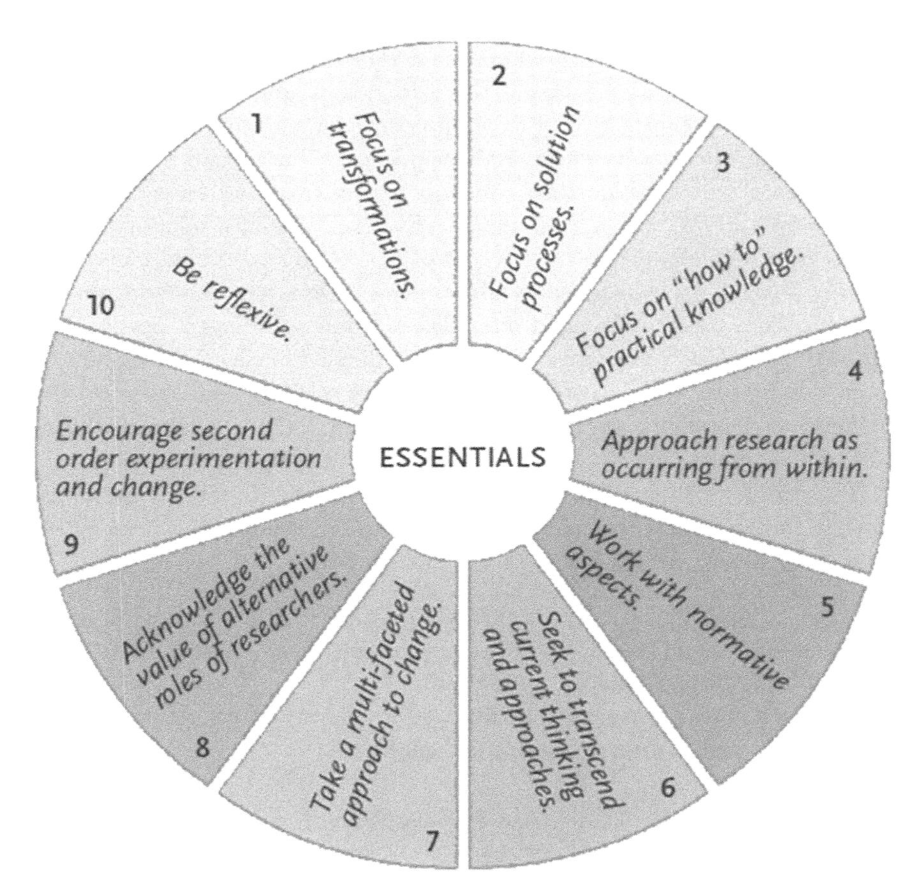

Figure 4.14 Ten fundamental principles for high-impact transdisciplinary research.

Source: Fazey et al. (2018, p. 60).

4.3.9 Transdisciplinary Co-Creation for Better Transformation Processes: Thomas Bruhn, Jeremias Herberg, Giulia Molinengo, Daniel Oppold, Dorota Stasiak and Patrizia Nanz (Germany, 2019)

The framework proposed by (2019) is intended 'to support policymakers in their efforts to address complex societal challenges within the context of a broad transformation towards sustainability' (p. 2). The authors advocated the idea that contemporary challenges can no longer be faced through the traditional and unidirectional policy model and therefore the structure is based on the integrated approach of reflection and co-creation of policy formulation in dynamic contexts characterized by a high degree of uncertainty and early involvement of decision makers and other stakeholders.

As regards co-creation,[16] this is the terminology given to processes that facilitate the development of a collective approach to a problem, while respecting some criteria. The most important is the opportunity that the situation provides for everyone to 'contribute their different perspectives and competences to the process and facilitate a joint and shared development of solutions' (p. 5). The main objective of the co-creative approach is 'to facilitate political decision-making through a form of collaborative process design that takes into account and integrates the broadest and most diverse range of relevant perspectives possible' (p. 7). Co-creative processes are instrumental in consolidating mutual trust, solving conflicts and strengthening the viability of joint solutions and understands that 'that none of the actors involved in the process are capable of understanding the problem alone, or indeed of identifying its "solution"' (p. 7).

The search for a deep and integrated understanding of a specific challenge in close cooperation with the affected people and institutions has been called Grounded Action Design and consists of four phases:

i. Problem Scoping: iterative process where all stakeholders affected by the problem and decision makers engage in dialogue to ensure that all aspects of the problem are considered.
ii. Transformative Mapping: participatory exploration of how (re)framing the problem will impact affected stakeholders.

---

16 The term co-creation was introduced by CK Prahalad and Venkat Ramaswamy in 2004 at the time of release of the book *The Future of Competition* (Prahalad, C. K. and Ramaswamy, V. (2004). *The Future of Competition: Co-creating Unique Value with Customers.* Boston: Harvard Business Press.) as a 'joint creation and evolution of value with stakeholding individuals, intensified and enacted through platforms of engagements, virtualized and emergent from ecosystems of capabilities, and actualized and embodied in domains of experiences, expanding wealth-welfare-wellbeing' (Ramaswamy and Ozcan 2014, p. 14).

iii. Identifying Stakeholder Capacities, Useful Ideas and Possibilities for Change: detailed analysis to identify the potential in terms of empowerment and expectation of transformative change.
iv. Developing Transformation Strategy: creation of a customized change process to address the complex challenge, based on the results from the previous phases.

Figure 4.15 shows how the phases of the Grounded Action Design are related.

The theoretical framework of Grounded Action Design is based on the tripartite and integrated vision of the following elements of the research project: (i) scientific analysis of co-creative policymaking processes, (ii) learning-oriented dialogue with relevant communities of practice and (iii) design of innovative prototypes for policy advice, whose area of intersection is the Co-Creative Policy Consultation. Figure 4.16 shows the tripartite vision of the project.

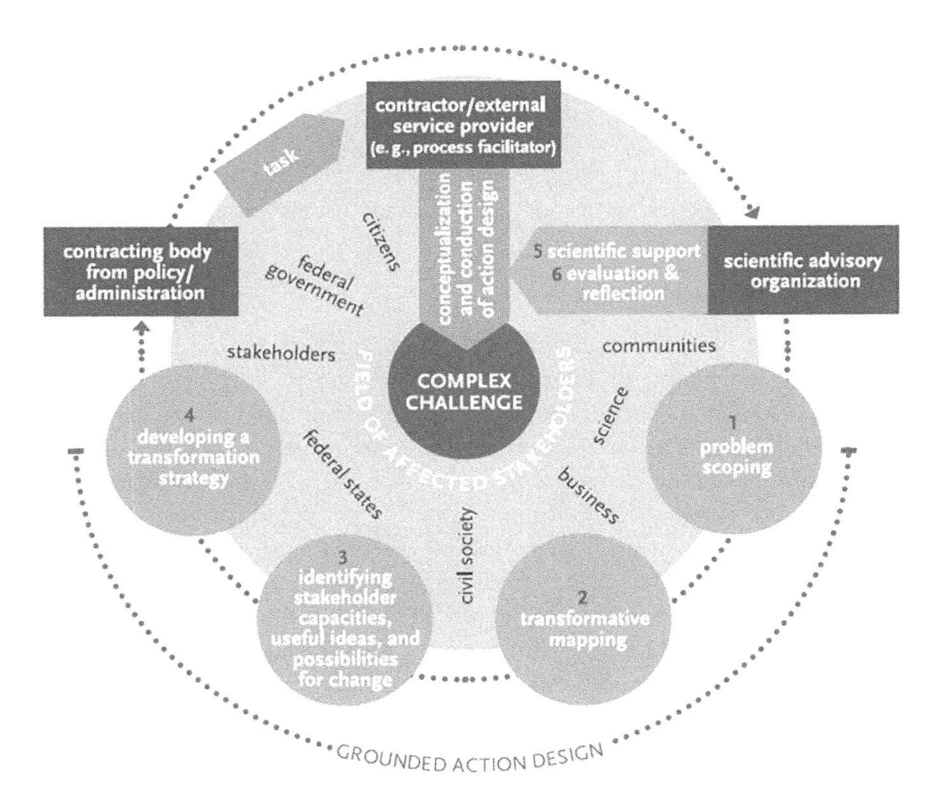

Figure 4.15 Action-based design: framework for a reflective and co-creative process.

Source: Bruhn et al. (2019, p. 8).

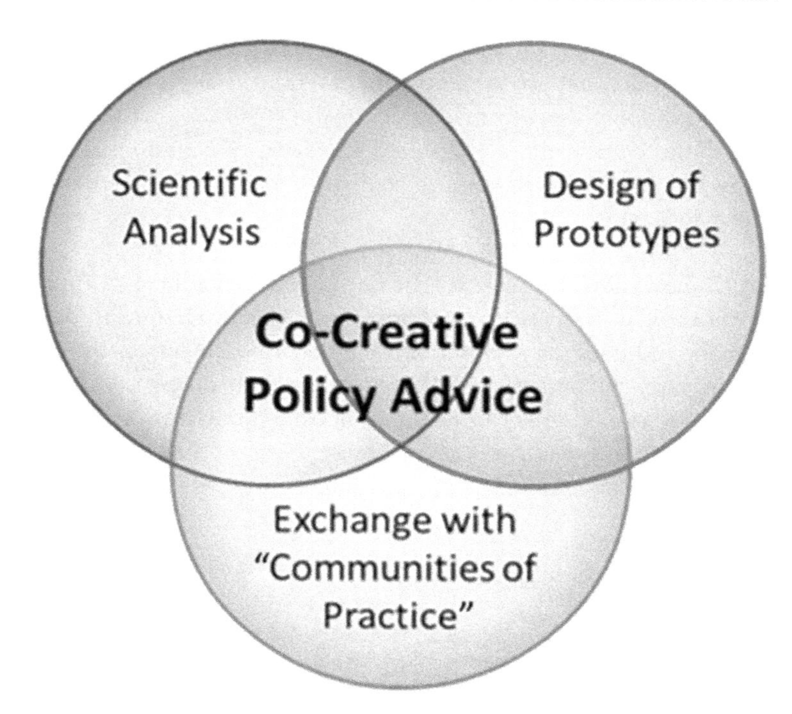

Figure 4.16 Central elements for co-creation of contemporary policies.

Source: Bruhn et al. (2019, p. 6).

4.4 Partial Considerations

Interdisciplinary and transdisciplinary research consists of integrative studies, which differ depending on the degree of interaction between their academic and non-academic participants. Interdisciplinary interactions refer to the intensity of cooperation and integration of the various disciplines in solving the research problem. Transdisciplinarity, in turn, adds the involvement of non-academic actors to this setting.

The main characteristic of integrative research is the union of various fields of knowledge and practical experiences, mitigating aspects of differentiation and specialization. Transdisciplinary research, in turn, coordinates different bodies of knowledge by identifying gaps in the science–technology–society tripod using appropriate scientific methodology and theoretical references.

The core of transdisciplinary research, therefore, is the search for solutions to a socially relevant issue that leads to the common good. Bacon pointed out, precisely, that science serves human progress and empowerment to achieve the common good. In the evolution of the transdisciplinarity *continuum*, Julie Klein explicitly stated the common good as characteristic of transdisciplinary research and indeed, this ideal has been an important preoccupation of philosophy, politics and law since ancient times.

Transdisciplinary research has vigorously advanced as a methodological framework for academic research. The conceptual frameworks are numerous and, for this book, the most significant one emerged in 2005, especially in Europe. The pioneering one was the model of the ISOE, while the most recent one, from 2019, also from Germany, explores transdisciplinary co-creation for better transformation processes. It is worth noting that the conceptual framework of 2013, entitled Applied Integrative Research, by Gabriele Bammer, actually proposes a new discipline, called Integration and Implementation Sciences (I2S), whose objective is to support applied and integrative research to promote the change of policies and practices on the basis of understanding, synthesis and management of knowledge.

The content of this chapter intended to present the pillars for the definition of transdisciplinary research. The ideas presented are translated into the following words: transdisciplinary research is an integrative research aimed at solving problems that are beyond the disciplinary, interdisciplinary or multidisciplinary reach and even beyond academia, based on systems and complex thinking, whose cross-fertilization between the various options methodological frameworks, the conceptual framework and the practices of the conditioned element can result in new scientific specialties, new forms of socialization and more vigorous institutions.

In other words, transdisciplinary research is a type of integrative scientific research, which promotes the encounter of academic knowledge and situated knowledge, in order for individuals to perceive essential traits of the problem, considering the communicative, socio-organizational and cognitive-epistemic dimensions, in a permanent pursuit of the unity of knowledge, in order to support the transformations required in society.

Chapter 5

KNOWLEDGE ACQUISITION DESIGN (KAD): A FRAMEWORK FOR TRANSDISCIPLINARY CO-PRODUCTION RESEARCH IN KNOWLEDGE GOVERNANCE AND ORGANIZATIONAL LEARNING

5.1 Governance of Knowledge (GovK) and of Organizational Learning (GovL)

Since the mid-twentieth century, knowledge and organizational learning have taken a leading role in public and private organizations, since their value has been acknowledged for building and maintaining competitive advantage and for effective implementation of innovation. Knowledge management monitors, facilitates and manages the processes of creation and internalization of knowledge at the levels of individuals, as well as the sharing, storage, dissemination and institutionalization at the level of groups and organizations, in order to help overcome knowledge-dependent organizational challenges and promote transformation. According to Freire et al. (2021):

> Organisational learning is a dynamic, systemic and continuous organisational macroprocess, which institutionalizes organisational knowledge that is created from four processes – intuition, interpretation, integration and institutionalization – at various organisational levels – individual, group and organisational – carried out on the basis of the tension between exploration and exploitation, in which *feed forward* occurs for assimilation of new learnings and feedback, for the use of what has already been learned. (p. 37)
>
> Knowledge management macro process (tacit, implicit and explicit) through the combination of sources (human and non-human), for the purpose of decision-making (strategic, tactical and operational),

> making use of the processes of auditing, acquisition, treatment, storage, sharing, disseminating and applying organizational knowledge to that significant knowledge (consolidated, under development and under construction) can be made available and used effectively, especially to generate learning transformational and organizational performance improvement. (p. 71)
>
> Organisational learning governance is the organisational system for the development of dynamic capacity and self-organisation, which governs collective cognitive and behavioral processes, through an interrelated set of mechanisms, components and learning environments for coping with and giving prompt response to changes. (p. 71)
>
> Knowledge governance is the organisational system composed of structures and a set of mechanisms, formal, informal and relational, to mitigate transaction costs and risks and transfer intra- and inter-organisational knowledge, established by corporate governance and knowledge management, to optimize organisational economic results. (p. 72)

As knowledge is the result of learning processes, that is, it is created and transformed by means of social interactions, the drivers of organizational transformation and innovation are dependent on governance structures, as these can enhance or block knowledge management processes, making it possible, or not, to achieve the intended results (Fleury and Oliveira Junior 2001; Freire et al. 2017).

The most recent studies on these relationships (Foss and Mahoney 2010; Pemsel and Müller 2012; Rizzatti and Freire 2020; Rizzatti 2020) argued that the mechanisms and components addressed by the traditional Corporate Governance model have blocked knowledge management and, consequently, organizational learning, making it difficult to carry out the transformation and innovation planned by the strategy of organizations. The traditional Corporate Governance model, currently in force in organisations, began to be questioned for having a restricted focus on monitoring and controlling results, and offering few mechanisms that support the cycles of knowledge and organizational learning, which promote innovation.

With simplicity, some authors such as Marques (2007, p. 13) summarized Corporate Governance as 'mechanisms or principles that govern the decision-making process within a company,'[1] whose main objective would be 'to balance the competitiveness and productivity of a company with responsible and transparent management'.

1 Our translation.

Although the need for monitoring and control cannot be refuted, we need to move forward to face the challenges posed by the collaborative economy and new intelligent technologies (which have been changing and improving more and more quickly); corporate governance models must be based on organizational mechanisms and components that recognize the value of sharing, exchanging and transferring knowledge for value creation, such as the multilevel governance model (Bobbio 2005; Piattoni 2009; Couto 2018; Freire, Kempner-Moreira and Hott-Jr 2020).

Organizational learning is a process by which organizations create, develop and institutionalize knowledge generated in their internal and external networks of cooperation and partnerships. Knowledge resulting from the learning process, in turn, is the basis for the future activities of individuals and groups within the organization, as well as influences how groups interpret their work environment, develop new insights, make decisions and solve problems (Crossan, Lane and White 1999; Vera and Crossan 2004). As mentioned previously, learning is the result of social relationships, that is, it depends on intra- and inter-organizational networks.[2] Networks efficiently leverage gains by (i) identifying problems and opportunities, producing flexible solutions that allow learning to be adjusted according to the complexity and variety of problems; in other words, networks favour proactive governance; (ii) sharing and gathering information and knowledge between participants, who become an asset in policy formulation; (iii) establishing a framework for building consensus or minimizing conflicts and; (iv) reducing the risk of participants' reluctance to the implementation of agreed policies (Pereira 2010, p. 9).

In this context, organizations, when creating and sharing knowledge and achieving organizational learning, become learning organizations (Senge 1990). Therefore, these organizations need to know how to govern their knowledge and organizational learning processes in order to develop and innovate on a continuous basis. As a result, governance of a series of objectives, types, mechanisms, components and environments is required to process multilevel learning as well as provide feedback, so that organizational objectives can be achieved.

In this way, for a contemporary organization to be capable of creating, acquiring and sharing knowledge, while modifying its behaviour to generate

2 An intra-organisational network is an organisational network internal to the organisation, composed only of internal actors, who interact, dialogue and co-create knowledge and knowledge assets, with the explicit objective of co-production of the common good, and they can be classified as networks for exchange, development, expansion, action and learning (Freire et al. 2021).

new knowledge and perceptions (Kiernan 1998), governance of knowledge and organizational learning is crucial, as it involves the governance of networks for collecting and sharing data and information from multiple actors for the creation of new knowledge (Nonaka and Takeuchi, 2004) through continuous learning, with the direct objective of delivering competent organizational performance.

For these contemporary organizations, and for universities whose research locus is knowledge-intensive organizations, the Framework Knowledge Acquisition Design (KAD) emerges as a conceptual framework for transdisciplinary research on co-production of scientific-technological knowledge, as characterized below.

5.2 Methodological Alignment

The KAD Framework is intended for applied research that uses a transdisciplinary co-production methodology, characterized by a robust theoretical foundation integrated with the organizational field. In compliance with the recommendations of Hadorn, Pohl and Bammer (2010), it is in line with one of the five strategies that are needed for integrative research: it is a framework based on dialogue, as it incorporates the communicative dimension in a dialogic and bidirectional dynamic from the moment of conception, passing through the moments of situation and creation, until reaching the transformation moment. With a qualitative or quali-quantitative approach, KAD supports propositional research (Alves-Mazzotti 2001; Bobbio 1997; Patton 1988) to advance scientific and technological knowledge.

To this end, through a more traditional view of research in the organizational field, KAD encompasses an exploratory stage (Creswell 2007; Gil 2009; Marconi and Lakatos 2009; Vergara 2005), a diagnosis of the organizational context (Roesch, 2009) and a descriptive stage of the constituent elements of the object of study to be analysed (Creswell 2007; Gil 2009; Marconi and Lakatos 2009; Vergara 2005), but its objective should be considered as propositional (Alves-Mazzotti 2001; Bobbio 1997; Patton 1988), just like action research.

As a result of its efforts, KAD promoted the creation of technological-scientific knowledge that is socially grounded to be accepted as generalized, as it is generated in academic and non-academic dialogue, dismantling the specialist/lay dichotomy, in accordance with Klein's prediction (2013).

The guidelines for KAD's application method, defined by similarity, are the characteristics of action research identified by Lewin (1946). However, there are also the characteristics proper of integrative research and the propositional objective that characterizes transdisciplinary co-production research. Thus, seven guidelines have been set out for the research methods based on the KAD Framework:

i. Be geared towards the organization's future and development.
ii. Include stakeholder representatives, promoting a participatory and collaborative environment.
iii. Establish a constructivist and co-productionist vision by seeking convergence and symmetry between the social and epistemic dimensions.
iv. Respect the procedures of each step of KAD to ensure compliance with the phases of transdisciplinary co-production research: review, inclusion, diagnosis, planning, co-creation, action, evaluation, application and learning.
v. Produce action-based theory.
vi. Recognize that the objectives, problem and method of research must be generated in co-production on the basis of the transdisciplinary research process itself.
vii. Propose solutions to general problems, starting from situational diving.

Based on these characteristics, it can be seen that the rationale behind it is pragmatism, a school of thought that considers practical consequences or real effects as vital components of meaning. KAD emphasizes that research makes a relevant contribution to the context of application, that is, the transdisciplinary effort is only justified if it is socially useful and takes the common good into account. As a framework that exemplifies this rationale, KAD is an 'overarching framework from which the selected problems and other similar problems should be approached' (Kockelmans 1979, p.72). Its efforts result in the creation of socially robust technological-scientific knowledge (Klein 2013) that can be accepted as generalized.

By encouraging co-production from the beginning of research studies, the results contain the least inference of the academic researcher, include the verification of the consistency of interrelationships and the validation of its application, and have the goals of uniqueness of knowledge and the interrelation of the elements that compose and influence the research object. Aligned with the claim of Ollaik and Ziller (2012, p. 229), studies that are predominantly qualitative , for example, those that are structured by KAD, should be concerned with being validated in the 'conceptions related to the stage of formulation of the study (prior validity), conceptions related to the stage of development of the research (internal validity), and conceptions related to this stage of the results of the study (external validity)'.

KAD seeks the unity of knowledge through an interdisciplinary view and uses transdisciplinary methodology, owing to its inherent pragmatism, to achieve the product for the common good. Moreover, the procedures require the inclusion of external actors in the process of research co-production. Accordingly, KAD supports the methodological structure of integrative research, which has participatory-collaborative characteristics and includes external actors. Aligned with the evolution of maturity stages of intra- and inter-organizational learning networks, established by the researchers Kempner-Moreira and Freire (2020), KAD respects the degrees of participation of actors in research projects, as proposed by Cornwall and Jewkes (1995). That is, research that uses the KAD methodological framework must always include actors inside and outside the academy; however, the relationship can be established in a contractual, consultative, collaborative or collegiate degree (typologies described in this book in the Participation and Collaboration section).

Further analysis of the patterns of definitions of transdisciplinarity identified by Pohl's study (2010) shows that the KAD Framework is based on the pattern that points to four possible situations in the characterization of transdisciplinarity: (i) to relate to relevant social issues, (ii) to transcend disciplinary paradigms, (iii) to do participatory and collaborative research and (iv) to search for a unity of knowledge. There is actually a fifth situation (v): to apply the result of the unity between technological and scientific knowledge.

KAD suggests integrative research as an approach to be followed in order to integrate academic and practical knowledge – from the stages of situation analysis, problematization, passing through the convergence and divergence of ideas for the acquisition and application of scientific and experiential knowledge and, finally, production of learning that can be expanded to innovation and transformation. The three dimensions of integration listed by Bergmann et al. (2012) stand out in KAD, and they direct the focus of the participants' attention while they experience the spaces for planning, problem, dialogue and solution. Thus, from the first conception moment, passing through the moments of situation and creation, until reaching the transformation moment, the communicative, socio-organizational and cognitive-epistemic dimensions are particularly relevant.

- The communicative dimension, which is established with a bidirectional and dialogic mission, will be required in every process, from the collection of data, information and knowledge, through the definitions of the project scope in the problem space environment, as well as in the discussions to generate ideas, develop prototypes and pilot test solutions, to the solution space and its outcome validation moments;
- The socio-organizational dimension is present in KAD's co-production proposal, recognizing not only that many actors should participate,

but also that all parties involved, whether academic or non-academic, public or private, should collaborate and

- The cognitive-epistemic dimension, when the framework requires six answers in the problem space and another six in the solution space, making twelve questions that direct the actors to listen to what others have to say about the phenomenon or study object. Whether these are human or non-human agents, such as documents and technological artefacts, they must be analysed and, subsequently, be subject to the cycle of consistency verification and validation by the participants involved.

In KAD, effectively, integration processes transcend knowledge integration and advance to multilevel learning. In the communicative dimension, a common discursive practice is intended, clarifying and creating terms and concepts. In the socio-organizational dimension, the objective is to understand the different interests and lead them to the intended outcomes, without ignoring the learning instantiated to the experience and interests of each participant. It is in the cognitive-epistemic dimension that the integration of knowledge is certainly expected, the connection of specialized knowledge bases, scientific foundation and foundation in social practices, in order to develop new knowledge, and enable the transfer of learning to practice, by building organizational methods and theoretical and methodological references.

By respecting the procedures of each step structured by KAD, one can fulfil the phases of a transdisciplinary co-production research study: the literature review, the inclusion of the actors, the situational diagnosis, the project planning, the co-creation of the solution, the research development action, the verification and evaluation of the results, the application and the expansive learning of the actors to the organizational practice. By the way:

> expansive learning is the learning process that goes beyond the environment where the educational action takes place, expanding to simulation of future practice in the workplace and, in the student's social practice, disrupting the encapsulation of theoretical knowledge, expanding it to include the relationships between the context of the educational institution and the context of practical application. (Freire et al. 2021)

Based on the analytical structure of Enengel et al. (2012), supported by the typologies of knowledge portrayed on the scales, in the functions and epistemologies proposed by the authors, KAD, in an almost linear logic, but not limited to it, deals with the scientific-technological knowledge, which will support problematization by having the multiples research actors hold discussions on the basis of empirical evidence or scientifically recognized theories, thus eliminating the discussions arising from individual beliefs. Precisely for this reason, a robust literature review is required, as well as a situational diagnosis of the organizational context.

Here, it is worth a deeper look into the literature of special interest to this work, called grey literature, which presents more specific characteristics than conventional literature, such as small print run, reduced audience, may have a provisional or preliminary character, without standardized ISSN or ISBN numbering and without peer review, among others. Typical examples of grey literature include reports, theses and dissertations, patents, standards, newsletters, blogs, videos and white literature[3].

The definitions of grey literature vary according to the time and the researcher's own point of view. Almeida (2000) sought the origin of the definition of the term and found that in the late 1970s it received greater meaning and became in current use from the seminar on the subject that took place in York (England), from December 13 to 14, 1978, organized by the British Library Lending Division. The term was adopted by English librarians, becoming popular in the 1980s. After the event, Campello, Cedón and Kremer (2000) highlight that a remarkable fact that occurred in relation to bibliographic control of European grey literature was the creation in 1980 of the System for Information on Grey Literature in Europe (born with the acronym SIGLE and later changed to OpenGrey), whose mission is to promote access and use of grey literature produced in Europe. It currently operates through an online, centralized and multidisciplinary database fed by 16 countries.

Población (1992) points out that another important event was the First International Conference on Grey Literature, held in 1993 in Amsterdam, whose results point out that 'conventional literature [...] does not match the speed required by changing societies [...] grey literature, being the unconventional, is dynamic and facilitates communication between scientists, administrators and communities [...]' (p. 245). The event maintains discussion about the meaning of the term and at its Third Conference, held in 1997, coined the structuring definition, 'that which is produced on all levels of government, academics, business and industry in print and electronic formats, but which is not controlled by commercial publishers' (National Grey Literature Collection, 1997). At the Twelfth Conference, held in 2010 in Prague, it noted the primary characteristic of the previous definitions and proposed:

> grey literature stands for manifold document types produced on all levels of government, academics, business and industry in print and

3 It is literature whose publication principle is commercial.

> electronic formats that are protected by intellectual property rights, of sufficient quality to be collected and preserved by library holdings or institutional repositories, but not controlled by commercial publishers i.e., where publishing is not the primary activity of the producing body." (National Grey Literature Collection, 2010).

It is important to note that grey literature is included in a specific type of literature review, the Multivocal Literature Review (MLR), which includes both academic and non-academic literature appropriate to the research topics or segments with a wide variety of sources and little scientific literature available. This typology of review recognizes the need for multiple voices rather than considering only rigorous, peer-reviewed scientific knowledge. Ogawa and Malen (1991, p. 265) define:

> Multivocal literatures are comprised of all accessible writings on a common, often contemporary topic. The writings embody the views or voices of diverse sets of authors [...]. The writings appear in a variety of forms. They reflect different purposes, perspectives, and information bases. They address different aspects of the topic and incorporate different research or nonresearch logics.

In a more structured way, grey literature is also incorporated into the Qualitative Systematic Review (QSR) protocol of Cochrane, the international non-profit organization that is the benchmark for systematic literature reviews. By the way, the purpose of the QSR is to bring the findings in documents from primary qualitative studies, such as ethnography, phenomenology, case studies, grounded theory studies, and qualitative process evaluations. Also included are studies that use qualitative methods for data collection, for example, focus groups, individual interviews, observation, document analysis, questionnaires, as well as qualitative methods for data analysis, for example, thematic analysis, structure analysis and discourse analysis. Studies that collect data using qualitative methods are excluded but analyse the data using quantitative methods, such as questionnaires where response data are analysed using descriptive statistics only (EPOC, 2020).

In the problem space, initially, we deal with specific knowledge of the context that is serving as a unit of observation and action. With this focus, one can identify and acquire the rich experiential knowledge, derived from life experience, often tacit or implicit to the practice of societal actors, achieving the motivation for sharing and contributing to research. In this context of diving into experience, local social and environmental phenomena and their descriptions diverge and converge, producing a type of knowledge that can be categorized as phenomenological. As a result, strategic knowledge is produced by the connections and interrelations of the elements studied and simulated, and finally, when expansive learning is established and knowledge is generalized, made universally valid and expressed in a systematic manner, free from context-specific conditions and restrictions, it is accepted by all parties involved.

According to Mitchell, Cordell and Fam (2015), KAD dialogues with outcome spaces, establishing the boundary of the project in terms of scope, resource and deadline, since there is no quality research without a feasibility assessment. However, KAD recognizes the constructivist process of co-production between theory and practice in search of solutions, while accepting the persistent uncertainty of the field. In this way, in light of the *Open Innovation*[4] model (for the porosity of inter-organizational and inter-institutional boundaries for the creation of learning networks), systems thinking in a project as promoted by Cabrera and Cabrera (2018) helps to promote the perspectives explored in Mitchell, Cordell and Fam (2015) Model and, consequently, in KAD.

Based on the conception of the boundary from Weyrauch's (2016) research project, KAD recognizes the value of inter- and intra-organizational interaction dimensions; they are both in the interdisciplinary vision of researchers and in their integrative dialogue with actors outside universities and academic research itself, but they are active in the co-creation and co-production of results and solution of the collective problem.

The 10 principles listed by Fazey et al. (2018) are respected in KAD, and they must receive the attention of researchers involved in studies based on it, as they must maintain the focus and transformation, solution processes and practical knowledge, bringing research into the solution development processes for, especially, transcending the limiting approaches traditionally employed. However, KAD privileges the eighth principle that acknowledges the value of the alternative role of researchers, stressing that researchers will

4 Open innovation refers to a flow of resources, knowledge and information that easily moves across the porous boundary between business and market. According to Henry William Chesbrough, who introduced the concept in 2003, it is a paradigm that assumes that companies can and should use both external and internal ideas, and either external or internal access to the market, so that they can advance their technology (Chesbrough 2003).

transcend the role of consultants, as stated by Bruhn et al. (2019), and become proactive in the problem space and be co-producers in the solution space.

KAD is also aligned with the vision of Bruhn et al. (2019) when it comes to the sequence of steps to be taken with a focus on driving action design. However, it reverses the logic proposed by this model (Figure 4.15), making step 5 (scientific support) as the first step to be taken, and the reflection and evaluation proposed by de Bruhn et al. (2019) will no longer be just a first stage, but as established by KAD, it will be transversal to the entire research process, and consistency and transdisciplinary validations will be verified in co-production among researchers, organizational specialists and societal actors.

Finally, the KAD Framework follows Pohl and Hardorn (2017) and makes explicit, in its definition, applies the common good as its main objective. As a definition, for the authors, KAD is a framework for designing and carrying out integrative research of transdisciplinary co-production in knowledge governance and organizational learning, which is structured through processes of action and transformation, whose guiding principle is the pursuit of the common good.

5.3 Rationale for Proposing KAD

The KAD is the proposal for a framework aimed at integrative research in the theoretical-practical body of knowledge governance and organizational learning, a critical segment of organizations, responsible for the interdisciplinary dialogue between corporate governance, knowledge management and organizational learning, in order to optimize the creation, use and transfer of organizational knowledge for development and innovation. The main conceptual frameworks discussed in this book, despite having made relevant and indispensable contributions to the debate on issues about transdisciplinary research, are not focused on the theme of KAD.

The ISOE Model for the Transdisciplinary Research Process is focused on sustainable development and on specific environmental issues, not delving into themes related to GovL and GovK The Principles for Designing Transdisciplinary Research compare to KAD in the emphasis on identification, structuring and analysis of the problem and fruition of the results, however, KAD aspires to reach all segments of scientific research, with the argument that conditioned knowledge needs unconditioned knowledge for reality to be understood and, unlike the principles put forward by Pohl and Hadorn (2017), it does not have the perspective of issuing guidelines only to transdisciplinary research practitioners. The Analytical Framework for Co-production of

Knowledge was developed to meet the needs of doctoral research, and it differs from KAD with regard to this limitation. Integrative Applied Research, in turn, encompasses a range that is not possible for KAD. It even proposes a new discipline, called Integration and Implementation Sciences (I2S) to support integrative applied research. The Transdisciplinary Outcome Spaces resemble KAD in highlighting moments of situation, knowledge and learning; however, it is limited to the options of the Soft Systems Methodology (SSM), based on Critical Systems Thinking (CST), to be able to evolve in context. The Four Building Blocks of Systems Thinking offer – as a conceptual framework – a analytical tool for practitioners to understand and solve complex problems, based on the foundations of systems thinking, and contextualization is the differentiating element of KAD. The Context of the Interaction between Research and Government Policy, according to the title, seeks to introduce the knowledge resulting from research in the context of public policies, limiting itself to this scope of action. The Ten Foundations for Contributing to Transformational Change are appropriate in KAD for the GovL and GovK approaches. And finally, the Transdisciplinary Co-creation Framework for Better Transformation Processes focuses on politics and sustainable development. Thus, the offer of this framework may cover this gap and enrich the understanding of the potential of transdisciplinarity in scientific and technological research.

KAD reinforces that the moral and philosophical core of the transdisciplinary effort resulting from a research project is the pursuit of the common good, followed by the guarantee of transformative learning, one that leaves a legacy and contributes to changing the situation. Its structure prescribes the parameters and guidelines of situated, experiential and expansive learning, emphasizing the need for actors to be fully aware of the broad meaning of research and how this influences their practice, learning and aspirations, as well as how its results will have a beneficial impact on everyone.

The framework completely rejects the participation of actors with a partial view of the research study, either because of participation for a fixed period of time, or because of imminent change among research members, or because of a high degree of specialization. In true transdisciplinary co-production, everyone needs to know the reality placed, the paths selected, the solutions adopted and the available infrastructure that shaped the context, objectives, methodology and expected results. This requirement is linked to the direct need for expansive, but also experiential learning, in which, by the way, ‘it is a continuous, circular, dynamic, holistic and multidimensional process of individual learning, carried out through the cycle of experiencing and transforming reflection on the own experience, in useful knowledge’ (Freire et al. 2021, p. 36). The inconstancy regarding the collaboration and participation of actors would make it impossible to promote dialogue, real co-production and, finally, the creation of target and transformational knowledge (Hadorn et al., 2008), blocking the transfer of knowledge to practice.

For each team member to uniquely contribute with their typology of knowledge and dimension for analysis of the object and phenomenon observed and experienced, making the project an expansive experience for each one's practice, the beginning is necessarily constituted by processes of sensitization and awareness about the use of their specific knowledge and the resulting mobilization for participation, collaboration and co-production.

Resuming the arguments of Mitchell, Cordell and Fam (2015), the transdisciplinary environment promoted by KAD leads to mutual and transformational learning for the actors involved in the system. However, this will not be established by traditional processes and practices, but rather by a new andragogical adult education method (Knowles 1980) for the organization that is the locus of research, which awakens experiential (Kolb 1984) and expansive (Engeström 1987) learning, that is, by itself, transdisciplinary environment leads participants to identify and reflect on their worldview, identify their obsolete knowledge and open themselves to new knowledge that may bring scientific or technological solutions to their practice. Here, the concept of neolearning is worth of mention: 'andragogical methodological platform of experiential and expansive teaching-learning[...] with a view to reducing distances between actors and establishing a process of co-creation of knowledge and readiness for the transfer of learning to practice' (Freire et al. 2021, p. 85). The result, in this way, is from everyone, to everyone and perceived by everyone, according to their respective interests.

5.4 Transdisciplinary Integration Spaces

The framework is based on four moments of transdisciplinary integration, distributed in two collaborative work phases.

The first, entitled Conception Moment, occurs in Phase I of KAD's application. At this stage, non-academic experts are invited to participate in order to enable the development of a scientific-technological research project plan that meets the common objectives of academic and non-academic participants, changing theoretical subjectivity into objectivity of the reality to be researched.

The second, Situation Moment, seeks to explore the context, carrying out the situational diagnosis, which will consolidate the previously explored problematization. As of this moment, the project has already been circumscribed in an integration methodology, the result of which is the gradual adoption of common terminology and attenuation of specific cultures of the actors involved, in favour of a common culture of that intersection. The result is the construction of the social problem in scientific expression and understanding, a moment when critical transdisciplinarity is achieved.

The third, Creation Moment, through transdisciplinary experience of divergence and convergence of ideas, results in new knowledge that will be applied in prototyping and piloting. The KAD approach seeks to transcend common and expected thinking and approaches, considering that contemporary settings are likely to be different from the settings and thoughts that gave rise to them. At this stage of the project, one needs to ensure the connectivity of the various knowledge bases in a collaborative and participatory process. The first result of this step is the compatibility of objectives between the various actors involved, in order to achieve harmony between credibility (scientific adequacy) and legitimacy (stakeholder experience). The last one is to ensure that knowledge can be interpreted, integrated, experienced and used by the greatest number of people; it is about the ability to share knowledge.

In the fourth, Transformation Moment, expansive learning is processed into practice, generating results that lead to innovation and to the common good. This phase begins with the identification of the solution and the effective implementation of the research results; the latter are understood as an intervention in a social system and, therefore, may require further research and interventions. Underlying the main results, in the KAD approach, it is expected that all actors, academic and non-academic ones, involved in the project, must be able to understand new perspectives, get to know new strategies, use new tools, and ultimately have a transformative learning experience, which leaves a legacy and contributes to personal, situational and organizational change.

Within this macroprocess, the generated or appropriated scientific and practical knowledge is integrated into continuous cycles of consistency verification and validation, that is, the above-mentioned moments are not a linear set of steps to be followed sequentially. On the contrary, the learning of a moment can affect what had been initially programmed for later moments; also, it can change the result of previous moments, leading to a permanent recursive behaviour at all stages of the research, especially in moments intended to identification of the problem, creation of solutions and effective application. As a result of this *modus operandi*, the knowledge present and resulting from the project can be perceived and used as experiential, phenomenological, strategic and generalizable knowledge. Figure 5.1 summarizes the elements of the KAD Framework and their interrelationships.

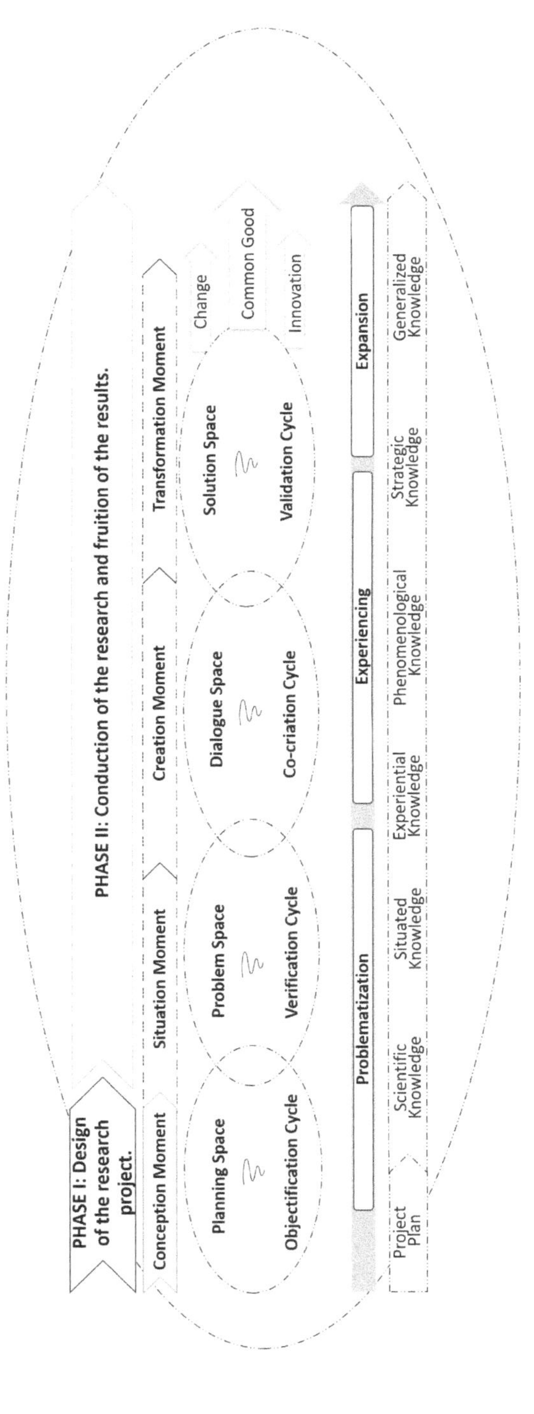

Figure 5.1 Constituent elements of the KAD framework and their interrelations.

Source: The authors (2021).

5.5 Instrumentalization

For reliability of the proposed instrumentation of KAD, in addition to its internal validation by the methodological alignment with the conceptual frameworks proposed by the preceding literature, the Design Science Research (DSR) is resumed, and new steps are added to respect the guidelines of the methodology of transdisciplinary co-production research.

Scientific research from the perspective of Design Science Research (DSR) is based on three cycles of achievement. The Relevance Cycle, which inserts the requirements of the contextual environment into research and introduces research artefacts into field testing; the Rigor Cycle, which provides theories and methods along with domain experience and background knowledge, and which adds the new knowledge generated by research to the growing knowledge base; and the Core Design Cycle, which supports a narrower cycle of research activity for the construction and evaluation of design artefacts and processes. Recognition of these three cycles in research design clearly positions and differentiates design science from other research paradigms (Hevner 2007). Although it is mostly used in information systems and production engineering, its *lato sensu* rationale is well suited to the broad applicability of transdisciplinary research.

Based on this perspective – and added to the critical phases of transdisciplinary research, integrative approach and governance of scientific and technological knowledge and organizational learning – the KAD Framework evolves through two phases. The first, divided into eighteen stages, is about designing scientific research, and the second is intended for carrying out the research and fruition the results among the participants.

5.5.1 KAD's Application Phase I: Designing the Research Project

This section describes the research Design Moment established by the collaborative planning space generated by the KAD Framework, distributed into stages of an objectification cycle, which in the end generates the project plan as a guide for the subsequent moments (Figure 5.2).

There are eighteen steps to design the project of transdisciplinary co-production research – initially established as sequential; however, respecting the constructivist and co-productionist view in the search for convergence and symmetry between the communicative, socio-organizational and cognitive-epistemic dimensions, each step can be revisited and redefined throughout the design process, as well as during the actual phase of the research for feedback on the plan that it is a part of.

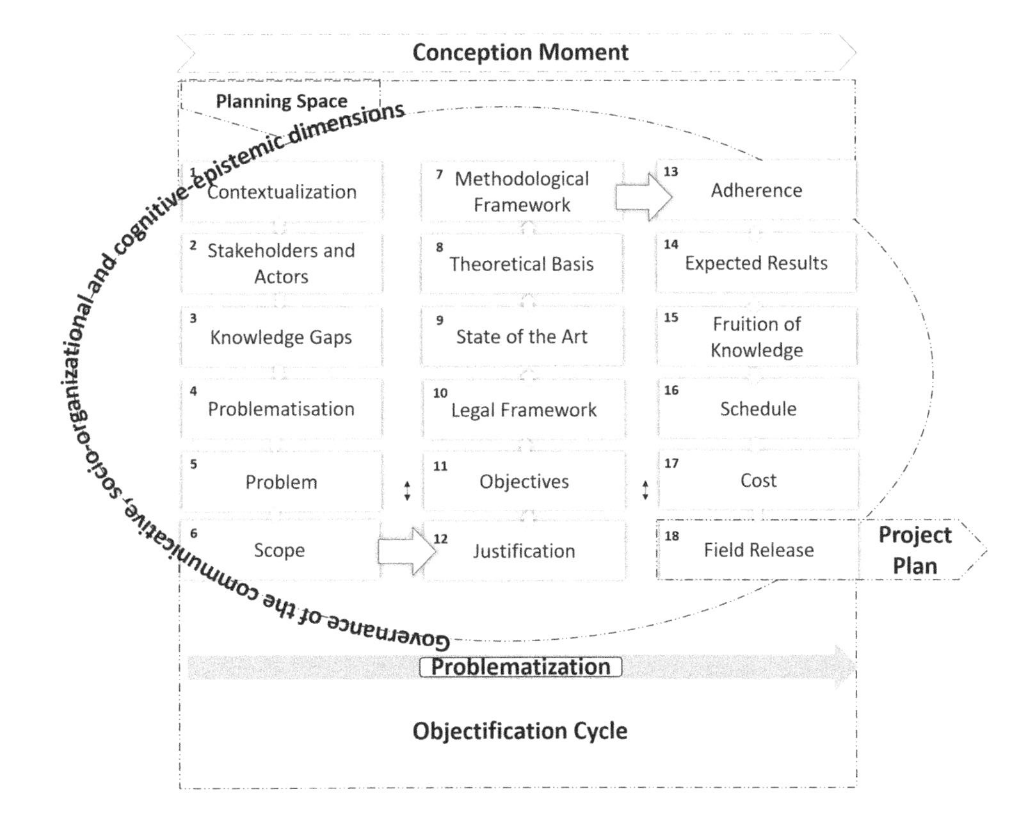

Figure 5.2 KAD's design moment and phase I.

Source: The authors (2021).

The eighteen steps that make up the KAD's Design Moment of Phase I can be understood as described below:

1. **Contextualization:** reveals the field setting[5] of the research, based on a literature review, including bibliometric data. Here, one can explore theoretical gaps, map the timeline of the main constituent milestones, distinguish publications and recurrent authors, identify topics of interest and understand the different and possible schools of thought.
2. **Identification of Actors:** specifies who the participants and collaborators of this research project will be and which academic and non-academic knowledge is essential to account for the complexity of the problem, search for answers and propose applied paths. Here begins the integration of all the constituent parts of the research project, which can be achieved through the creation of the project's collaborative group. It is advisable to define its social identity and positionality.
3. **Knowledge Gaps:** point out the paths for carrying out research, based on diagnostic field research that should include, in addition to previously explored scientific knowledge, situated, experiential and organizational practice knowledge. At this stage, one can see how the phenomenon one is researching occurs in the field, based on document analysis, interviews, questionnaires, focus groups with leaders or collaborators. The main procedure is to scrutinize the field, promoting a theoretical-empirical discussion that consolidates a situational diagnosis.
4. **Problematization:** explains the problem situation involving the study subject, based on empirical-theoretical discussions and on knowledge gaps. This step is useful for showing that complex problems are rarely solved in a single project. When problematizing, before the problem is identified, the expectation of closure of the problem question is definitely removed, since in the KAD approach each project promotes an improvement following the ideal that small changes, step by step, project by project, can lead to transformation.
5. **Problem:** explains the specific topic of the problematic situation that will be dealt with. The research question will guide the definition

5 Considering the importance of the context for applied transdisciplinary co-production research, the definition of field, in the context of the KAD Framework, is in order: place where the study phenomenon actually occurs, on the basis of the precise limitation of the object of study and the areas of science that dialogue to shed light on the organizational phenomenon and that were the object of research.

of the objective and the search for knowledge in the literature and in the field. In this phase, one can identify the intensity of cooperation and the need for integration of interested parties in search of a solution to the research problem, because a transdisciplinary context is needed to take the issue from the real world to a scientific research question. Here, as well, one establishes the disciplines needed to achieve results.

6. **Scope Definition:** delimits the object and phenomenon of study regarding the aspects of temporality and location in an iterative process where all parties engage in dialogue to ensure that all aspects of the problem are considered and that there is no disappointment throughout and at the end of the research.
7. **Justification:** briefly exposes the importance of carrying out the project, both for science and for the field, emphasizing the stage in which the situation is, the contributions that the research will bring, the importance of the project from a general point of view and for particular cases. This item, above all, needs to describe the common good to be obtained, that is, based on the socially relevant question pointed out during problematization, what the solidarity concern is with the members of that context that changes the way of thinking and acting of the project's actors, based on the transformational learning achieved for a conduct that can reflect the fact that the interests of others are their own interests.
8. **Goals:** define the project's path through specific objectives in order to achieve the general objective, which, in turn, answers the research question, within the expected scope. They clarify what changes will be affected as the research study is carried out.
9. **Legal Framework:** recognizes the legal basis that affects the project, considering that it will be limited by rules and regulations, sometimes complex ones, which may make the implementation of results and even the research itself unfeasible. If one identifies instructions provided for in law that make the implementation of the results unfeasible, they must be incorporated into the final guidelines, with suggestions and a detailed relevant review.
10. **State of art:** identifies the frontiers of knowledge of the researched theme, on the basis of bibliographic and documentary exploratory research. Here, will be more productive for the systematic[6] or integrative

6 The most important guide for carrying out a systematic literature review is the *Cochrane Handbook for Systematic Reviews of Interventions*, developed especially for the field of health, but whose methodology is widely used in all areas of knowledge. The Manual, *roughly speaking*, advises on review planning, search and selection of studies, data collection, risk of bias assessment, statistical analysis and interpretation

review, to support the arguments and identify how much the results can contribute to new discussions. As a result, the authors who have studied the topic, as well as the contribution of each one of them, the fields of application, the validated methods and proposals and their conclusions will be known. This step is contained in the literature review and the degree of robustness will impact the confidence regarding the originality and quality of the project's scientific and technological results. If carried out adequately, the bibliographic reviews that are planned, systematic and explained in this phase of KAD, establish prior verification as an alignment with what was determined by Ollaik and Ziller (2012) regarding the validation process in the conceptions related to research formulation.

11. **Theoretical basis:** establishes which of the theories and which literature will be vital to the foundation of the research. It differs from the broad literature review in that it is derived from it and brings an objective, precise and specific theoretical construction to the research segments. It is the construction of the scientific meaning of the research, the theoretical choice that will guide the research in action, and the support for transferring the results to practice.
12. **Methodological Framework:** justifies the choice for transdisciplinary co-production and describes the research design based on the steps of the KAD Framework. It explains the scientific method (preferably phenomenological), the nature[7], the way of approaching the problem (qualitative, quanti-qualitative), the type of research (preferably propositional with exploratory, diagnostic and descriptive steps), the technical procedures (preferably case study, action research[8] or participatory research[9]), collection techniques (interviews, questionnaires and focus groups) and data analysis to achieve each of the objectives, including how the interaction

---

of results. It is fully aligned with transdisciplinary co-production research, even if it does not use this terminology, as it is based, among others, on the following principles: (i) review authors should carefully consider which questions are important to different research stakeholders, as systematic reviews are more likely to be relevant, promote mutual learning, improve acceptance of results and reduce waste of time and resources; (ii) mapping all research stakeholders is the first step of the review; (iii) the Cochrane Declaration of Principles, which changes the research culture by bringing stakeholders together as joint research partners. (Higgins and Thomas 2019).

7 Basic research is not the object of transdisciplinary research. The object of theoretical and methodological references is applied research.

8 In some countries, transdisciplinary research is called action research.

9 Transdisciplinary research is necessarily participatory and eventually collaborative research.

spaces and the co-production of the problem, application, solution and learning will be developed. Other requirements are the transdisciplinary integration strategies used to get to this point and to move on to KAD's phase II. It is worth emphasizing even though transdisciplinary research is predominantly qualitative, it does not dispense with predicting the steps of verifying the consistency of the proposal and validating the results in the field. Finally, at this stage, there are also issues about research ethics; sometimes, the authorization of an ethics committee of the organization or of the academy is required in order for a study to move forward.

13. **Adherence:** relates the context and intended objectives to the chosen line of research and the field's strategy.
14. **Expected results:** estimates the answers, intended solutions and repercussions of the research, in addition to its theoretical and practical contributions arising from transdisciplinary integration.
15. **Fruition of Knowledge:** describes how everyone can be sensitized, made aware of and mobilized to apply the results achieved, taking advantage of the knowledge produced, shared and acquired as well as the desired common good. At this stage, the fundamentals of multilevel learning in organizations (of individuals, work groups and organizations) and the Knowledge Management processes are taken into account, and the methods, techniques and tools to be used to transfer the learning produced by research for the practice of organizations, are also considered.
16. **Schedule:** identifies the specific actions to be carried out throughout the project and the respective start and end dates for each activity. At this stage, one can identify the relationship of dependence between the activities and the critical path to enable the future management of the project, in an agile and collaborative manner, with the distribution of responsibilities between different academic and non-academic participants.
17. **Cost:** establishes the estimated costs of the research, determines its budget and explains the physical-financial schedule for the planned stages. Here, one can include the resources and infrastructure required for the research to develop properly.
18. **Field Release:** At the end of this journey, the transdisciplinary co-production research project will be designed in an integrated and collaborative manner, and the research itself can be started. For the latter to be developed with real involvement of non-academic researchers, and even to possibly enable free access to data and to people with knowledge about the organization, it is essential that the project be formally presented to the leaders of the partner organization in order to formalize its beginning and institutionalization of the collaborative group of researchers.

At the end of the course of the 18 stages, these must be described in a Project Plan document distributed among the actors directly involved in the execution of the research project to enable them to recognize that the objectives, problem and research method must be generated in co-production, departing from the transdisciplinary research process itself.

5.5.2 KAD'S Application phase II: Conduction of the Research and Fruition of the Results

The second phase of KAD application is organized into the three other stages of integration: Situation Moment, Creation Moment and Transformation and Learning Moment.

5.5.2.1 Situation Moment

The integration spaces for Knowledge Acquisition Design in phase II start at the Situation Moment. At this stage, the research begins with an in-depth analysis of studies on the research object and the effort to understand it in the organizational context. It is the time to collect the documents about the object or phenomenon of study, identify the domain specialists (people from the organization that have knowledge about the object) and the future users of the knowledge to be produced by the research. Next, in the problem space, the following questions need to be answered: what do stakeholders expect? what do the documents say? what do the actors need? what do the stakeholders complement? what do the experts validate, and what do the users apply? To answer the questions, data, information and knowledge of scientific and practical origin are needed. At the end of the Situation Moment, one must check the consistency of the resulting analyses.

The dynamic and propagating core of decisions and actions in the Situation Moment shapes the structure of the problem defined in KAD's Phase I. It is worth noting the specificity of the adjective 'dynamic', which means that repeatedly going back to the problem and the objectives, in a recursive cycle of action and evaluation, is part of the nature of transdisciplinary research, as transdisciplinarity refers to the animated state, to diligence; it carries with it the ability to handle problems on the go.

Implicit in this understanding is the certainty that the problems are not within a specific disciplinary framework and that, in addition to potentially needing an interdisciplinary context, they lack the involvement of practitioners, that is, the holders of situated knowledge that will actually emphasize joint responsibility for making decisions during the research. In summary, aligning the problem, objectives, scope and expectations right from the beginning of the research is a strategic resource to guarantee good results from co-production.

In the problem space, the following questions need to be answered: what do the stakeholders expect? What do the documents tell us? what do the actors need? what do the stakeholders complement? what do the experts validate? what do the users apply? (Figure 5.3).

To answer the questions of the situation moment, whose input of data, information and knowledge is significant, we need definitions that go far beyond the classic understanding that data, information and knowledge are in a sequential order in which each one is a constituent part of the next.

Figure 5.3 Situation moment of KAD's phase II.

Source: The authors (2021).

In this discourse, it is important to stop and explain what we, authors from the fields of information science and knowledge management, understand about the interdisciplinary relationship that is established on constructs.

Zins (2007) introduced the concepts in two domains. In the subjective domain, data is the sensory stimuli that we perceive through the senses; information is the meaning of these sensory stimuli; and knowledge is the individual's thought, characterized by the belief that the information is true or appears to be true. In the objective domain, they are human artefacts. Data can be words, numbers, digital signs, light beams, sound waves among others, that represent stimuli or perceptions; information is a set of the above signs that carry meaning and knowledge is a set of signs that have meaning and that individuals believe to be true.

From the philosophical perspective of Floridi (2015), data, pure data or proto-epistemic data are 'fractures in the fabric of being', revealed before being interpreted epistemically. They can be postulated as an external information anchor and can present themselves based on the following three independent characteristics. From the perspective of Taxonomic Neutrality, they refer to the fact that there are no data *per se*. Data are a property external to something and a relational entity, always associated with some context. From the perspective of Typological Neutrality, they are related to the fact that information can consist of different types of related data, of which five classifications are quite common, not mutually exclusive: primary data (the main data, directed primarily at the user, being 'what is in question'); secondary data (they are the inverse of primary data, constituted by their absence, through observation of what is not present, what is missing, what has escaped perception. In fact, this is a specific feature of information; its absence can also be informative); metadata (indications of the nature of other data – usually primary – that describe their properties such as location, format, update, availability, usage restrictions and so on. Correspondingly, metadata are information about the nature of information); operational data (relating to the operations and performance of a system. Correspondingly, operational information is information about the dynamics of a system) and derived data (those that can be extracted from other data, normally used in search of patterns, clues, evidence or inferences about things other than those directly addressed by the data itself, for example, for comparative and quantitative analyses). From the perspective of Ontological Neutrality, there can be no information without data, and information cannot possibly exist without physical implementation. And from the perspective of Genetic Neutrality, there can be *'information without an informed subject', that is, the* data that are present in the environment carry environmental information, which can have their own semantics regardless of a *producer*, known as *environmental data, which support* environmental information.

Despite Shannon's manifestation that '[i]t is not expected that a single concept of information would satisfactorily explain the countless possible applications of this general field' (Shannon [1993], p. 180), this book will adopt the explanation that data are the material from which the information is made and, therefore, information consists of one or more pieces of *data*, well-formed and *significant* or potentially significant and interpretable. And if well-formed and meaningful, the result is that there is *semantic content* in information (Bar-Hillel and Carnap, 1953). And if there is a great deal of semantic information, it can be said that there is a high degree of informativity (and vice versa). As a consequence, these notions are grouped around two central properties: information is extensive and information reduces uncertainty.

Information is extensive; its nature is to be broad, generic, comprehensive. It can be applied to an infinite number of people or things. It can be further extended; it is extensible. And it produces extension wherever it reaches, adding to what already exists. The notion of extension naturally emerges in interactions with the world around us when we absorb surrounding information. Information reduces uncertainty, the amount of information we obtain directly reduces uncertainty until such time as we get as much information as possible and the amount of uncertainty is zero.

To these two properties, one can also add reproducibility (information is reproducible without limits; it is reproducible); transmissibility (information is potentially transmissible or communicable; it is transmissible); it is quantifiable (the linguistic, numerical or graphic coding is quantitatively measurable or valuable; it can be quantified); it is perceived (what is called pregnancy, the founding and shaping action of information); it is the result of the interaction of the internal and external environment (which is called dynamic integration, the informational act as a result of both internal and external conditions and circumstances of the subject of the action). Or in Wiener's words:

> Information is a name for the content of what is exchanged with the outer world as we adjust to it... The process of receiving and of using information is the process of our adjusting to the contingencies of the outer environment, and of our living effectively within that environment. The needs and complexity of modern life make greater demands on this process of information than ever before. [...] To live effectively is to live with adequate information. Thus, communication and control belong to the essence of man's inner life, even as they belong to his life in society. (p. 268)

In order to introduce knowledge, we consider Zins' analysis; both the subjective and objective dimensions highlight the value of belief in truth to distinguish information from knowledge. There are, therefore, three components from traditional analysis that are present in the concept of knowledge: (i) it is true, (ii) it is believed to be true, (iii) it is justified because it is believed to be true. And with that we close the digression on the principles that govern the theme, whose depths are abundant, multiple and complex. This introductory paragraph aims not to ignore the tradition in knowledge studies; however, for KAD, the authors chose to go straight to the framework's typology of knowledge of interest, starting with scientific knowledge (or academic knowledge) and situated knowledge.

Scientific knowledge as such is the belief that it is true, it is produced reliably because it is based on solid foundations. What sets it apart from non-scientific knowledge is that it is produced through scientific research, meeting demanding epistemic standards, and that, as a result, it is highly reliable, robust, or well-established (McCain and Kampourakis, 2020). It is commonly associated with the best available knowledge, but this is not always necessarily so. It is best to associate it with reliable knowledge about a certain subject at a given time, since science, even if it is not infallible, is the most reliable way to arrive at non-obvious or non-superficial truth, through explicit evidence, data, observations, analyses and inferences. It requires being able to access evidence and understand how it relates to a research question.

The epistemology of scientific knowledge is founded on the principle that a great deal of contemporary science is collaborative, that is, scientific knowledge is collective knowledge, often obtained from groups – not just individuals. Cooperation (or co-production) is not an accidental feature of today's science; many questions are so broad and complex that answering them requires teamwork, at both practical and cognitive levels. At a practical level, even if a person had all the knowledge and skills, they would not be able to complete it without a collaborative setup. At the cognitive level, many research projects require multi- or interdisciplinary experience, skills and knowledge.

Regarding situated knowledge, it is worth clarifying the use of the term in its two distinct origins: the former, with a long twentieth-century tradition, refers to studies on knowledge from the margins of society, of unconventional science, understood as embodied and socially, culturally, racially, sexually, linguistically and politically situated, while the latter refers to learning studies that occur in the process of observation and engaged practice, closely linked to studies of tacit knowledge. Today, they have both been united in the studies of knowledge deriving from practice (Hunter 2009).

Based on this perspective, it is worth understanding exactly what empirical knowledge means as a situated activity. We can start this explanation with the definition of (Gherardi, 2008) about what practice is: 'It's a topos that connects "knowing" with "doing". It conveys the image of materiality, fabrication, of handiwork' (p. 517). The same author points out that the epistemology of situated practices for knowledge production has some important implications for organizations; the main one is the fact that there is no unified theory of practice-based knowledge. Of course, there are three non-excluding relationships, established between practices and knowledge: (i) a containment relationship, in the sense that knowledge is a process that takes place within situated practices; (ii) a relationship of mutual constitution, in the sense that the activities of knowing and practicing are not two distinct and separate phenomena, that is, they interact and produce each other; (iii) an equivalence relationship, in the sense that practicing is knowing in practice, whether the subject is aware of it or not. In all of them, knowledge is formed in and through action itself.

> We may say that in order to make knowledge observable in its making and un-making we shall focus on working practices as the locus of knowledge production, and reproduction; we shall pay attention to the dynamics between practice as institutionalized knowledge and practising as institutionalizing process, and we shall assume that knowing in practice is synonymous of practising. (p. 518)

When considering the knowledge process integrated to work practice, the term situated has a multiplicity of meanings. It can be understood, as pointed out by Hunter (2009) in this document, as *situated in the body*, related to feminist criticism, often the basis for specific competences, with emphasis on showing that even universal knowledge is situated in the body; it is anchored in the body. It can be *situated in the dynamics of interactions*, concerning the nature of the knowledge present in the interactions between humans and of humans with non-humans, as pointed out by Latour (1987). It can be *situated in language*, regarding the fact that all expressions change their meanings according to the subject and according to the context of use and, therefore, knowledge is produced through language or, as the author prefers to say, discursive practices, in order to emphasize that situating speaking practices is doing, regardless of the subjects who do that. And it can be *situated in a physical context*, regarding the fact that a workplace is a 'situational territory' (Goffman 1971); 'the materiality of situations enters into relations, objects can be conceived as materializations of knowledge, as tangible knowledge which "steers" and sustains a set of practices' (Gherardi, 2008, p. 521).

Within the framework of knowledge originated in practice, as an empirical and observable phenomenon, the concept of knowledge situated in an organizational structure is present in the relations established in

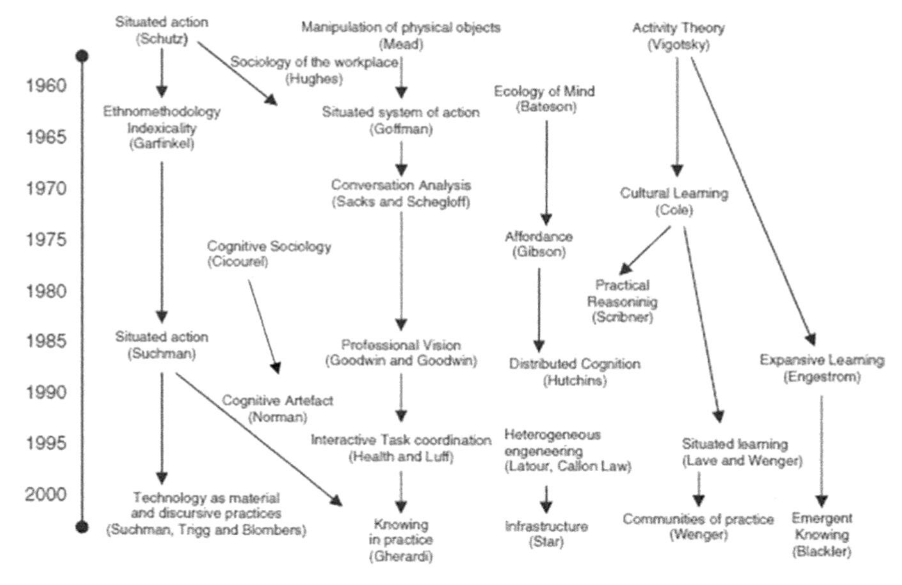

Figure 5.4 Flux of milestones on practice-based knowledge studies.

Source: Gherardi (2008, p. 522).

the community of practitioners that constantly produce and reproduce it, in a reconciled language, in a situational territory by means of institutionalized or institutionalizing practices, in a paradigm of co-production and co-creation. Here, it is worth presenting the flux of milestones on practice-based knowledge studies, which, in this work, highlight situated knowledge and expansive learning (Figure 5.4).

Back to the description of the Situation Moment of KAD's application phase II for the execution of the research and the fruition of the knowledge resulting from the transdisciplinary co-production, it is worth noting that, at the end of these first stages of Phase II, the research is verified internally, in line with what was determined by Ollaik and Ziller (2012) regarding the conceptions relative to the research development phase (internal validity).

5.5.2.2 *Creation Moment*

After collection of all the data, information and knowledge, in an intra-group collaborative research dialogue, the Creation Moment begins with the formation of the dialogue space. In a cycle of co-creation between academics and non-academics, there are two distinguishing features of this second Moment: (i) acquisition and integration and (ii) application and experimentation, both in a dynamic and feedbackable context of divergence, convergence, prototyping and piloting.

This is the stage of generating ideas, testing and seeking consensus on the solution to be created, through a series of debates and dynamics of co-creation between the participants, in order to clarify specialist knowledge and the different interpretations of academics and non-academics on the partial results achieved so far. The individual, explained interpretations are now combined, integrated and placed at the group level, giving rise to the design of possible solutions to be tested in practice. The second phase is the process of application and experiencing, which results in experiential knowledge and phenomenological knowledge. This phase consolidates the prototyping and piloting of solutions, whether at the stage of theoretical models or of minimum viable products to be simulated and tested in practice, and they dynamically feedback the transformations of updated versions of the results.

The Creation Moment follows from the previous moment, adding the dialogue space to the transdisciplinary co-creation process, as shown in Figure 5.5.

We will now pause the presentation of KAD and talk about these three essential concepts to the dialogue space: acquisition, integration and application of knowledge, because for KAD and certainly for all participants of transdisciplinary co-production research, whether they are individuals or organizations, processing knowledge from external sources expands the knowledge base, shaping these three vital knowledge management processes.

Knowledge acquisition refers to the ability of an organization to identify and acquire knowledge produced externally or internally, as regards critical knowledge for its operations, which in turn refers to the:

> knowledge needed to guide the action that will lead to global or partial organisational results. It is directly associated with the epistemological issue of the value of knowledge; it is valuable because of the role it plays, which makes it essential to examine the relevance of its purpose and, therefore, identify the critical factors of knowledge in an organisation…, relevant knowledge (innovative content, with technical content appropriate to the strategy) and vulnerable knowledge (of difficult acquisition and training, difficult to capture and transfer in context, and scarce).[10] (Freire, Alvares, & Silva 2021)

According to the open innovation paradigm, which adopts an integrative perspective when considering internal and external sources of knowledge,

10 Our translation.

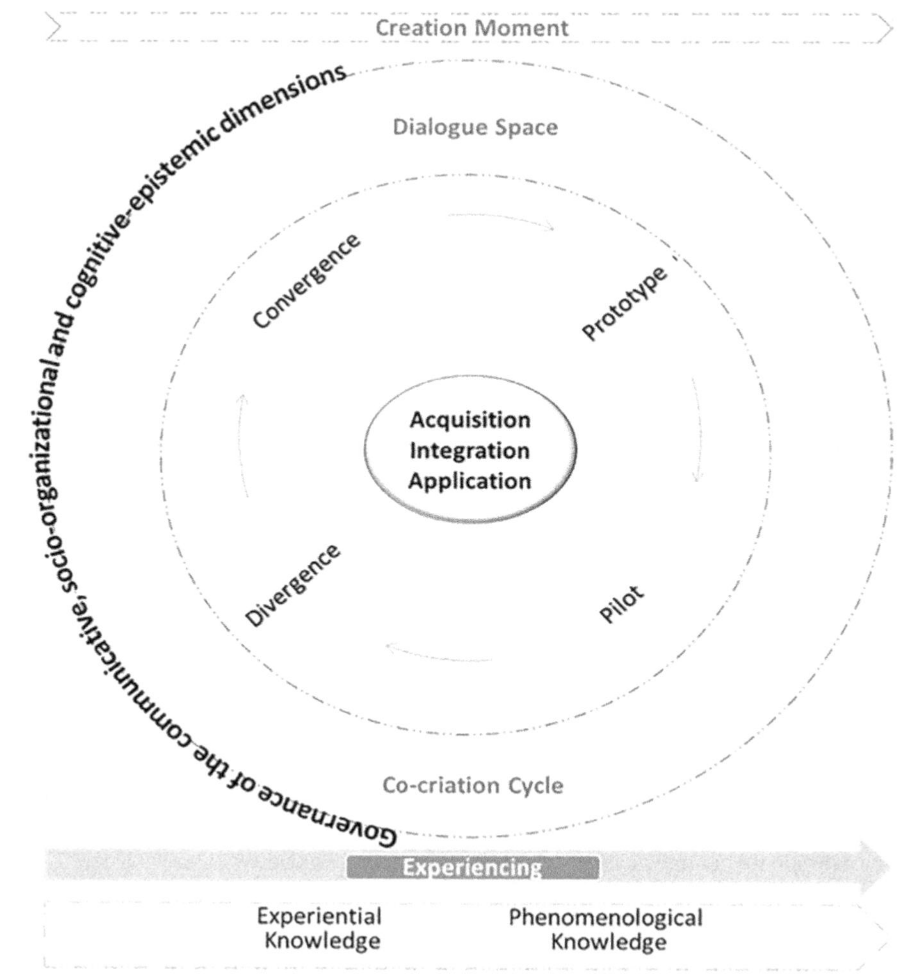

Figure 5.5 Creation moment of KAD's phase II.

Source: The authors (2021).

organizations must acquire knowledge from external actors to integrate with internally developed knowledge, which is in line with West and Bogers (2014), who presented the issue as a summon to people within organizations to seek external knowledge and integrate it with internal knowledge to improve processes and products.

Papa, Dezi, Gregori, Mueller and Miglietta (2018) went further; they argued that knowledge acquisition is indeed one of the main aspects of the open innovation paradigm, since most innovations today are more dependent on knowledge acquisition than on internal knowledge creation, which leads to collaborative arrangements and knowledge-sharing networks. It can be noted that knowledge acquisition is directly related to organizational learning: on the one hand, knowledge is a result of learning and, on the other hand, knowledge is one of the inputs to the learning process. For Liao, Chang, Hu and Yueh (2012), learning, at the organizational level, is the sum of all knowledge-related processes, which can be grouped into knowledge acquisition (KA) and knowledge management (KM). Moreover, in Huber's (1991) view, knowledge acquisition, along with information distribution, information interpretation and organizational memory, are the four constructs of organizational learning and specifically refer to how knowledge can be acquired through talent, accumulation of experiences, guided learning, transfer and search of knowledge.

The integration process concerns the capacity to absorb knowledge, when new knowledge is added to the previous knowledge base. At an intermediate stage, adding new knowledge requires learning, especially when it is beyond existing capabilities. Integration is also necessary to absorb the specialized knowledge of the research group members, which is equally indispensable; it results in the combination and integration of existing knowledge with acquired knowledge, which is sufficient to reset existing organizational capabilities.

Zahra and George (2002) translated absorptive capacity as a dynamic capacity consisting of two components: potential absorptive capacity, which comprises the acquisition and understanding of external knowledge, and realized absorptive capacity, which is the internalization, conversion and use of knowledge. This classification leads to two other insights: the transfer of knowledge and, if successful, the use of the transferred knowledge.

One should show any possible barriers to the use of transferred knowledge, as compiled by Bierly, Damanpour and Santoro (2009), such as resistance to change, inability to assimilate externally generated ideas, lack of effective knowledge-sharing techniques, and differences between external knowledge and the company's knowledge bases.

The effective ability to acquire and integrate knowledge does not mean that it is being used to achieve previously defined goals. In fact, the application, that is, the use of critical knowledge, is the main objective of knowledge management; its value derives from its application. Chen and Huang (2009) argued that when organizations are able to use relevant knowledge, they reduce the likelihood of making mistakes, decrease redundancy, increase efficiency, and continually translate their organizational expertise into embedded products. Shujahat, Sousa, Hussain, Nawaz, Wang and Umer (2019) pointed

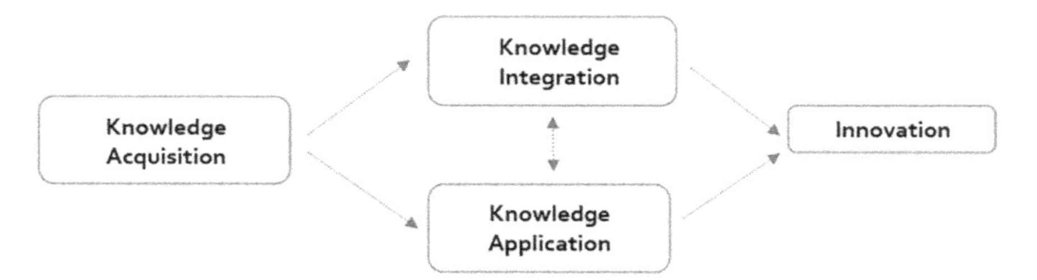

Figure 5.6 Knowledge processes directly related to innovation.

Source: The authors (2021).

out that the application of knowledge is more important than other knowledge management processes, because knowledge is not important until it is applied for the benefit of an organization. Boateng and Agyemang (2015) described the application of knowledge as processes within organizations that enable them to use and leverage knowledge in ways that improve their operations, develop new products, and generate new sets of knowledge.

Figure 5.6 shows the processes of acquisition, integration and application of knowledge that, in KAD, are portrayed as directly related to innovation.

In a one-sentence summary of the three knowledge management processes highlighted in KAD, the main objective of knowledge management in a transdisciplinary co-production study is the application of knowledge obtained from the integration of critical pieces of knowledge acquired from sources that are internal and external to the study. And for this process to occur, experimentation is necessary.

The experience of the Creation Moment of transdisciplinary co-production research aims to establish a continuous, circular, dynamic, holistic and multidimensional process of collective learning, carried out through the experimentation cycle (prototyping and piloting) and transformation of the reflection on one's own experience (divergence and convergence) into useful knowledge for research and organizational reality.

The term convergence used in KAD borrows its meaning from biology, when referring to the phenomenon of 'convergent evolution' of the ideas explained by the research participants. In the dialogue space, similar characteristics are identified in the ideas, even if they originate in the theory brought by researchers or in cases of practice experienced by non-academic specialists. The same holds true for the term divergence, when referring to the phenomenon of evolutionary divergence, equally borrowed by KAD from biology. In this case, it is a question of identifying divergent ideas among the participants of the same academic, disciplinary or organizational origin, and then disseminate the adaptation of their peers to the new ideas generated during the transdisciplinary dialogue.

For innovation, divergence and convergence are caused by active techniques and methodologies of group dynamics, determining an understanding also recognized by KAD. In summary, in the Creation Moment, diverging thoughts must be encouraged to generate a large number of ideas. Then, the convergence of thoughts must be encouraged to refine the countless suggestions, focusing on the good ideas that can be transformed into solutions.

Prototyping refers to the design process, followed by the construction of the project, and the evaluation by the research group and the future user. This process can be restarted as many times as necessary to improve the transformation and to add value to the resulting knowledge. The prototype is a rudimentary representation of what the application of knowledge will become to enable a detailed examination of its general operation. In other words, even if primitive, it is a representation of the idea to discuss and analyse future solutions and possibilities. It enables the active interaction of the composing elements and provides the opportunity for initial application tests, offering a broad and real view of the functions (Sperandio and Évora, 2005).

The prototype, therefore, can be seen as a representation, not necessarily functional, of an idea still in the design stage. The prototyping proposed by KAD should be understood as the creative space to explore ideas generated in the research before time and resources continue to be invested in its development or implementation. The use of prototypes during transdisciplinary co-production provides the consolidation of knowledge relative to the domain of the problem in question, because when the project team presents the prototype of an idea, even if primitive – to discuss and analyze solutions – its objectives, constituent elements and fundamentals are collectively understood. In prototyping, converging and diverging research ideas are encouraged to identify and promote improvements.

Prototyping is followed by piloting, an activity planned as a test in an environment very similar to the reality in which the research results will be used. For KAD, piloting is a real, temporary, and exclusive testing process for collective assessment and learning. Convergence and divergence of ideas are encouraged, but this time with the direct participation of organizational experts and future users of the knowledge produced. During piloting, one can include additional resources, use key people and adjust the budget and plans that had been originally established. It requires careful monitoring, because it is based on the evaluation and learning of the pilot project that the new one will be founded and will begin to be used in real projects.

In the Creation Moment, experiential knowledge is present, understood as the 'knowledge asset generated by the tacit knowledge of experts, created from experience, resulting from the exercise of social and work activities'[11] (Freire et al. 2021, p. 55) and the phenomenological knowledge, based on Edmund Husserl's phenomenology, which recognized that knowledge is limited to the interpretation of phenomena by human consciousness, whose philosophical understanding of phenomenon is the way in which something manifests itself to the subject on the basis of its social reality, enabling the assumption that all knowledge is also knowledge of itself and is constructed from countless perspectives of human consciousness.

Together, the Situation and Creation Moments, as shown in Figure 5.7, carry out a cycle of external verification of qualitative or predominantly

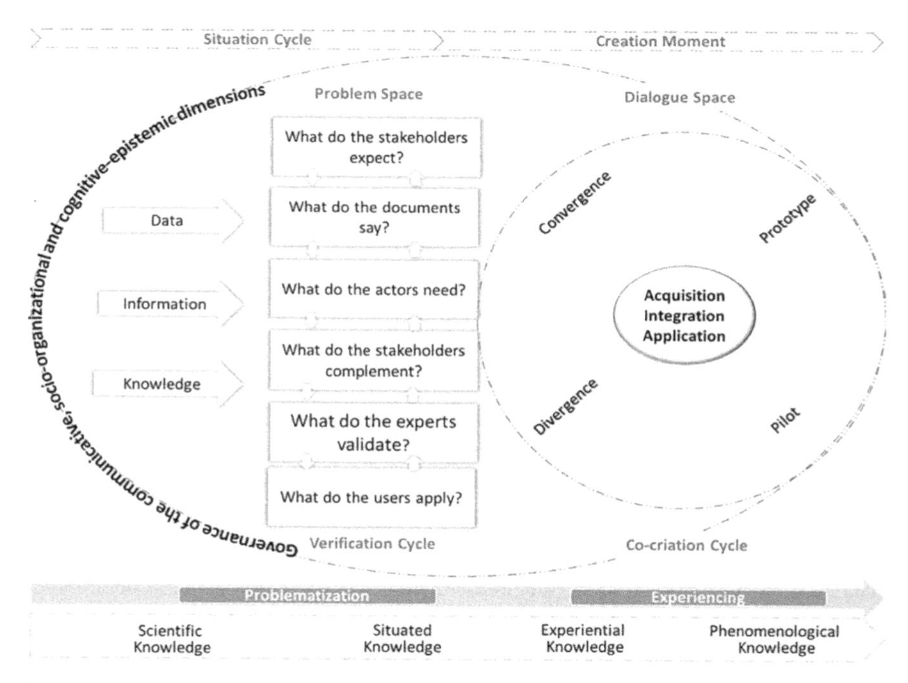

Figure 5.7 Situation and creation moments of KAD's phase II.

Source: The authors (2021).

11 Our translation.

qualitative research (Ollaik and Ziller, 2012), on the basis of the governance of the communicative, socio-organizational and cognitive-epistemic dimensions.

Now is the time to identify:

- Interacting external forces[12]: pointing out facilitators and barriers arising from the structural and circumstantial factors of politics, economy, history and culture that establish the research development scenario. It is essential to identify them in order to achieve successful results, on the one hand and, on the other hand, make them ready to support the expectations of research actors, as well as, if necessary, refer back to bibliographic reviews with the aim of identifying studies that support the choice of the way forward, taking into account these interacting external forces.
- Interacting internal forces: pointing out the elements that will be rigorously present alongside a research study: culture, organizational capacity and management, and essential resources (budget, time, infrastructure and technology, subdivided into other critical dimensions). The potential for change will be closely related to each of these elements and, therefore, should be a structuring part of the research scenario.
- Information and knowledge infrastructure: revealing the current strategies for information and knowledge management, in order to ensure that the research results can be not only appropriated by the structure of knowledge and organizational intelligence, but also be a permanent reference for the project.

5.5.2.3 Transformation Moment

The transformation moment, the passage from one state to another, is characterized by the expansive learning process and begins when the following questions are answered in the solution space: what the critical practical and academic types of knowledge are, what their meanings are, how they are constituted, how they interrelate, how to represent them and how to apply them, in a process that, as a whole, has resulted in innovation, change and the common good. Figure 5.8 shows the relationship of the constituent elements of the transformation moment.

---

12 Here, a terminological clarification is needed: research participants are also called actors and the other elements that interact with the research are the agents, individuals or organizations affected by the research and by the actors' actions.

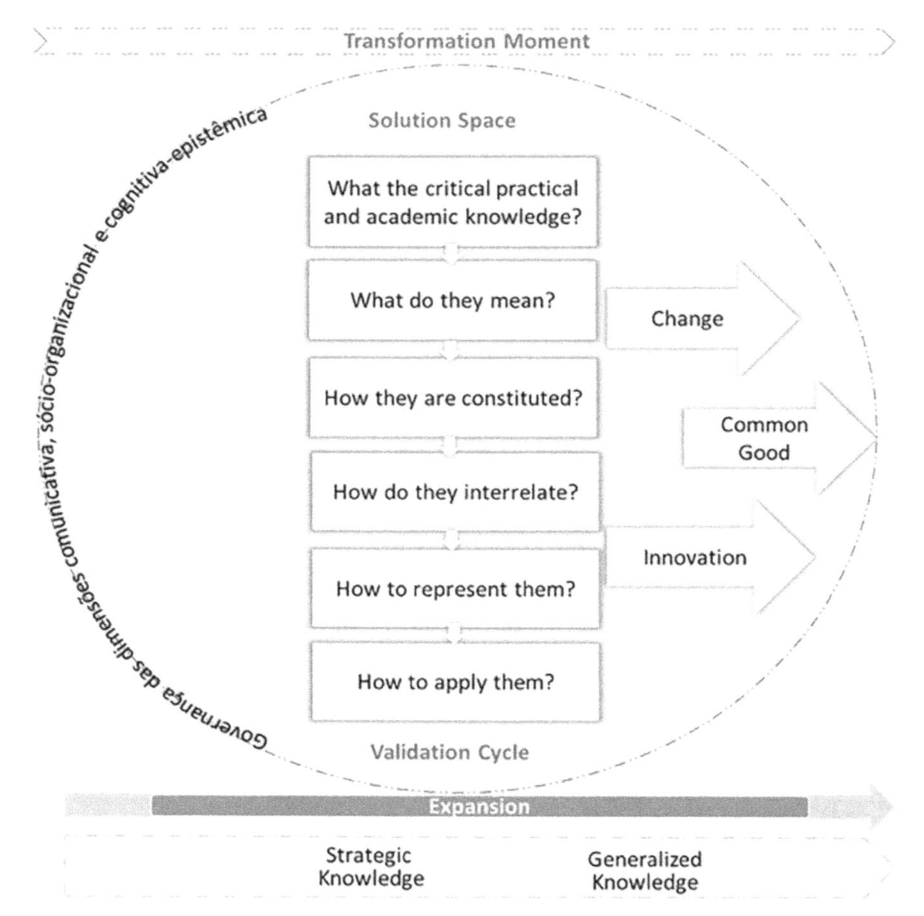

Figure 5.8 Transformation moment of KAD's phase II.

Source: The authors (2021).

Now is the time to identify:

- the unknowns: managing the unknown surrounding the result, arising from uncertainties, imprecisions, mistakes, distortions, incompleteness, suppressions, taboos, resistance, lack of resources and knowledge, elements of reality.
- possible improvements: helping teams to understand how the government, organizations and civil society operate and how the results will fit with them, validating, or not, the theories that underpinned the research. The same occurs for the opposite path; the research results are input for the government, organizations and civil society.

In order for the research propositions to be defined, the validation of the results achieved so far must be processed in the organizational context that served as the field of study and by specialists external to the collaborative research group.

At this moment, all conditions are ready to enable (i) change, (ii) innovation and (iii) the common good. The expected change can occur in the strategic, structural, technological, human, cultural or political perspectives. As for innovation, improvements can be as much expected as disruptive innovations, but admittedly generated from open innovation through learning networks between academic and non-academic actors. However, all changes and innovations should privilege the common good that is related to participating organizations – academic and non-academic – and to the advancement of scientific knowledge.

The increase in knowledge is expected in strategic knowledge – a typology of organizational knowledge related to planning and strategy that seeks to align resources and capabilities to achieve competitive advantage, composed of both the explicit knowledge and the implicit and tacit knowledge of decision makers – and in the generalized knowledge – universally valid, free from context-specific aspects, conditions and restrictions, with low cost of dissemination. However, its application generally requires translation and adaptation.

Figure 5.9 shows the interacting elements of Phase II of the KAD Framework.

5.6 Partial Considerations

The main objective of this book is to introduce the KAD framework, that is, to provide a detailed conceptual framework for transdisciplinary co-production research in knowledge governance and organizational learning. The description begins with the concepts related to the organizational field that is object of the framework, organizational learning, knowledge management, organizational learning governance and knowledge governance.

The characterization proceeds with the methodological alignment of transdisciplinary co-production, bringing what is advocated both by integrative participatory research and by the fundamentals of transdisciplinarity in its relationship with the search for the unity of knowledge. Pragmatism is present in the philosophical context and the common good – the heart of transdisciplinary research – in the moral context.

The choices that are made can be supported by seven guidelines, the explanation of four possible situations in the characterization of

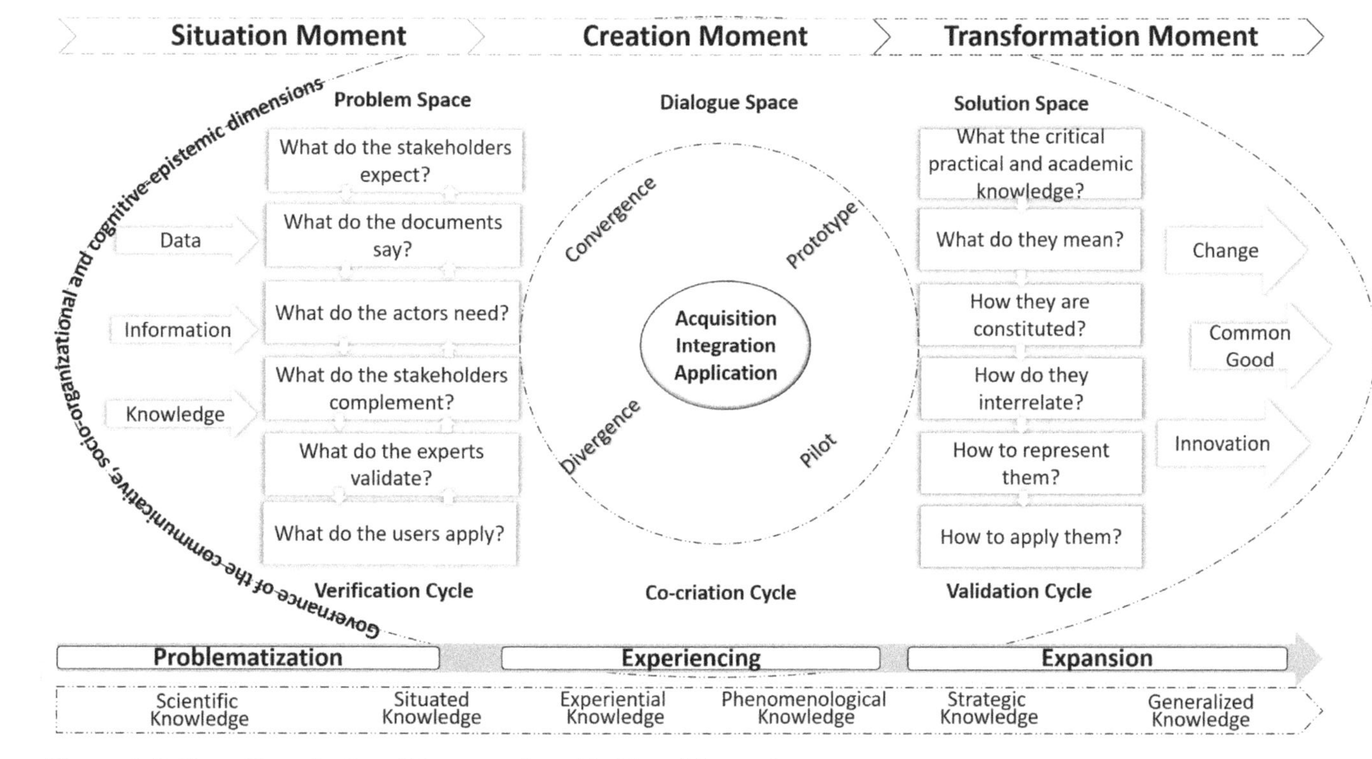

Figure 5.9 Phase II conduction of the research and fruition of the results.

Source: The authors (2021).

transdisciplinarity, the three dimensions of integration listed by Bergmann et al. (2012), Enengel et al. (2012), the distinction of transdisciplinary outcome spaces of Mitchell, Cordell and Fam (2015), the boundaries of inter- and intra-organizational interaction by Weyrauch, Echt and Suliman (2016), the 10 principles listed by Fazey et al. (2018), the subsequent steps for conducting research by Bruhn et al. (2019) and finally, the three cycles of achievement from the perspective of Design Science Research (DSR).

KAD instrumentation takes place in two phases. Phase 1 is intended for the conception moment of the research project, organized into eighteen stages, to enable the design of a project plan that meets the common objectives of academic and non-academic participants. Phase 2 is aimed at carrying out the research and fruition of the results, organized into three integration moments. It also contains details of concepts that are essential to the full use of KAD: data, information and knowledge; the perception of some typologies of knowledge underlying the organizational context; knowledge acquisition, integration and application processes; and divergence, convergence, prototyping and piloting cycles.

The content of this chapter was intended to introduce the pillars for proposition of the KAD Framework, destined to the applied research of transdisciplinary co-production methodology, characterized by a robust theoretical foundation integrated with the organizational field.

The result of KAD's efforts is the creation of scientific-technological knowledge, which is socially structured in order to be accepted as generalized, as it is generated in both academic and non-academic dialogue. The KAD Integration Environment bridges disciplinary divisions and encourages dialogue between academic and non-academic actors. In line with the axioms of transdisciplinarity, KAD realizes and highlights the multiple realities in conducting research based on scientific knowledge and situated knowledge.

In other words, Knowledge Acquisition Design is the framework for designing and carrying out integrative research on transdisciplinary co-production in knowledge governance and organizational learning, which is structured through processes of action and transformation contained in four moments and spaces of integration (conception, situation, creation and solution), based on the communicative, socio-organizational and cognitive-epistemic dimensions, whose guiding principle is the pursuit of the common good.

To conclude this work, the KAD Framework is summarized in Figure 5.10.

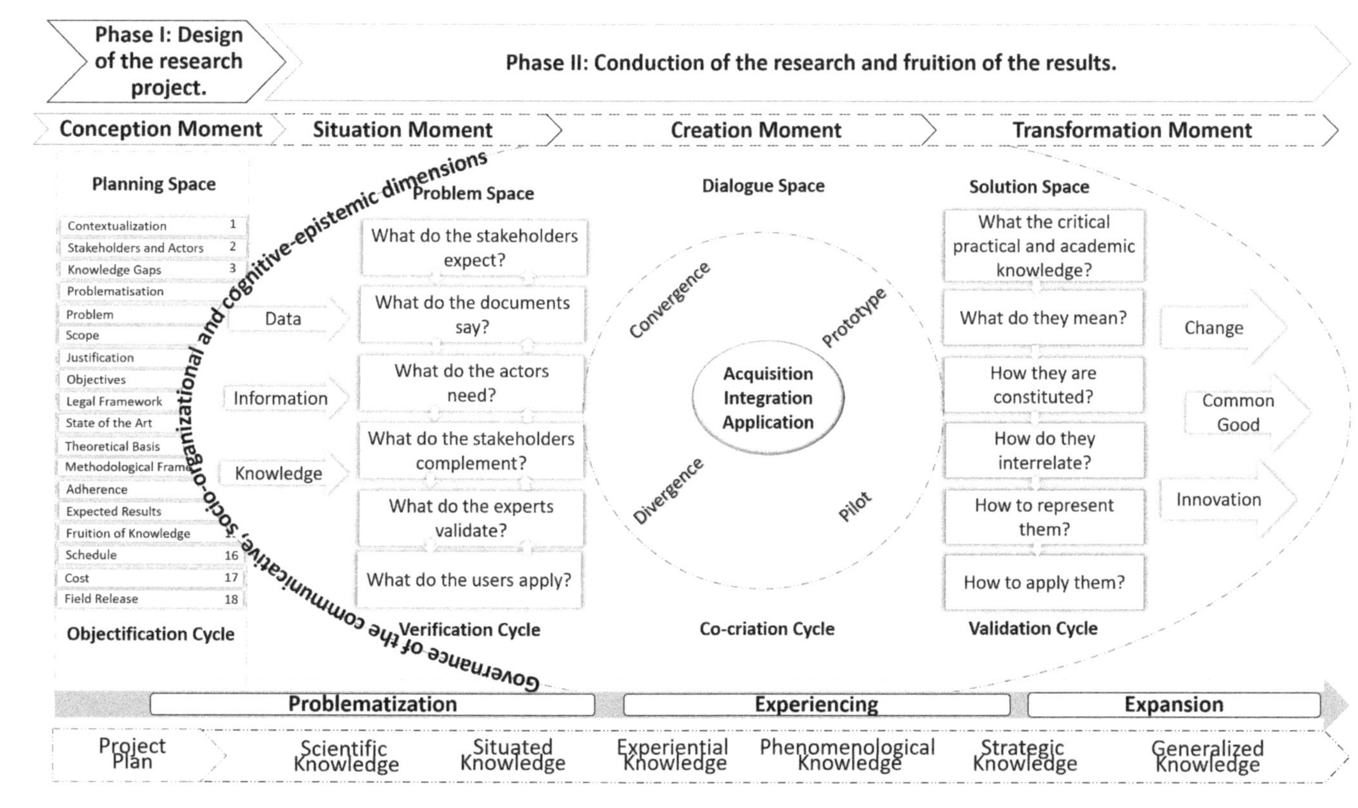

Figure 5.10 Knowledge Acquisition Design framework (KAD).

Source: The authors (2021).

Chapter 6

FINAL REMARKS

This book intended to discuss the fundamentals of transdisciplinarity and the concept of transdisciplinary co-production research, and propose the Knowledge Acquisition Design Framework, aimed at integrative research in the theoretical-practical body of knowledge governance and organizational learning (KLGov), in compliance with the conceptual framework of knowledge governance, an important construct for organizations that learn and innovate.

Throughout the chapters, some contributions and theoretical choices are noteworthy. The first is the use of the term co-production, which can take on several specific meanings; here, it is based on Sheila Jasanoff's studies of science and technology in 2004. The author described the co-production of scientific knowledge with society. She discussed the issue in more detail and pointed out that a significant goal of co-production is to generate knowledge on the basis of governance practices and to shed light on how governance practices influence the production and use of knowledge. With the meaning given by the author, the issue is qualified as transdisciplinarity of co-production or transdisciplinary co-production, with the research studies of Robert Frodeman (2014) and Merrit Polk (2015). The second important contribution, by Polk, is the perception that situated knowledge and scientific knowledge share the responsibility in the search for the solution to relevant problems. The author defined transdisciplinary co-production research, derived from integration processes for knowledge creation, as the combination of scientific perspectives with other types of relevant perspectives, which include co-production from the joint formulation of problems to quality control after implementation.

Another very relevant contribution of this book to the researchers who have read it is the description of the conceptual frameworks, as they were useful for defining the KAD framework itself. The frameworks are rich in detail, which makes it difficult to choose the most suitable one, or the one that will be the basis of a framework for research in the organizational field.

Each one is suitable for a specific situation; all of them have in-depth and well-defined theoretical ties and an excellent methodological basis.

The contributions of this book extend to the definitions of unity of knowledge, transdisciplinarity, transdisciplinary co-production, and transdisciplinary research, as shown in Table 6.1.

These definitions were developed in response to the challenges of transdisciplinary co-production research on knowledge governance and organizational learning; they may as well be suitable for other contexts. Indeed, based on the fundamentals examined in the four preceding chapters, the most important contribution of this book is the presentation of the Knowledge Acquisition Design (KAD) framework, which instrumentalizes the transdisciplinary co-production research for organizational studies

Table 6.1 Major Concepts of the KAD Framework

Term	**Definition**
Unity of knowledge	Consilient knowledge, whose unity can only be achieved through conditioned knowledge and unconditioned knowledge, the latter not only a contingent aggregate, but an indispensable system for the complete identity of conditioned knowledge.
Transdisciplinarity	Unity of knowledge beyond disciplinary boundaries, aimed at capturing the full complexity of the multidimensional and multireferential Reality of the conditioned element, by addressing relevant social issues, providing deep levels of transformation in higher education with a view to the co-production of scientific knowledge geared towards the common good.
Transdisciplinary co-production	Knowledge production in the confluence of academic and non-academic perspectives, in all phases of the process of inclusion, collaboration, integration, usability and reflexivity between the actors involved, co-creating with society and overcoming disciplinary, scientific and technological divisions, based on a fertile field of dialogue of the multiple realities of science and social practices, involving both the scientific method and the context where the problem occurs.
Transdisciplinary research	Typology of integrative scientific research, which promotes the convergence of academic knowledge and situated knowledge, in order to reveal essential traits that point to the problem, considering the communicative, socio-organizational and cognitive-epistemic dimensions required for integration, in search for the unity of knowledge, with the objective of supporting the so much needed transformations in society.

Table 6.1 (*Continued*)

Term	**Definition**
Organizational learning	Dynamic, systemic and continuous organizational macroprocess, which institutionalizes the organizational knowledge that is created from four processes – intuition, interpretation, integration and institutionalization – at various organizational levels – individual, group and organizational – carried out from the tension between exploration and exploitation, where *feedforward* occurs with the assimilation and projection of new learning, and feedback takes place for the use and improvement of what has already been learned.
Knowledge management	Management for the use and combination of (human and non-human) sources and types of knowledge (tacit, implicit and explicit) in (strategic, tactical and operational) decision-making through a set of processes for audit, acquisition, treatment, storage, sharing, dissemination and application of organizational knowledge to add value to products (goods and services).
Organizational learning governance	Organizational system for the development of dynamic capacity and self-organization, which governs collective cognitive and behavioural processes, through an interrelated set of mechanisms, components and learning environments for coping with and promptly responding to changes.
Knowledge governance	Organizational system composed of structures and a set of formal, informal and relational mechanisms to mitigate transaction costs and risks and transfer intra- and inter-organizational knowledge , established by corporate governance and by knowledge management, to optimize organizational economic results.

Source: The authors (2021).

related to knowledge governance and organizational learning in two phases of application (research project and research execution).

The first phase is the moment of designing the research; it consists of eighteen steps whose aim is to integrate the theoretical and practical knowledge of academic and non-academic actors in a planning space for the development of a viable project to be carried out and which meets the expectations of results for the common good. The second phase is composed of three distinct, interacting and interdependent moments for conduction of research and fruition of the results: the situation moment, the creation moment and the transformation moment that, respectively and dynamically, are integrated in the spaces of the problem, of the dialogue and of the solution.

In addition to contributing to the advancement of organizational studies based on a robust framework for transdisciplinary co-production research, the proposition of KAD consolidates the typologies of knowledge that must be considered and respected for the dialogue between academic and non-academic experts. Table 6.2 shows the respective definitions, based on the Glossary of multilevel governance of knowledge and learning and its mechanisms of networked corporate university and dialogic communication by Freire et al. (2021).

We conclude the presentation of KAD, summarizing its identity as the framework for the design and implementation of transdisciplinary integrative research in knowledge governance and organizational learning, which is structured through processes of action and transformation contained in three moments of co-production (situation, creation and solution). It is established

Table 6.2 Typologies of knowledge

Term	Description for KAD
Scientific knowledge	Typology of knowledge considered to be true by science, produced in a transparent and traceable manner, based on solid foundations, acquired through scientific research, meeting demanding epistemic standards, whose results are reliable, robust and well-established on a given subject at a given time.
Situated knowledge	Typology of knowledge located in an organizational structure present in the relations established in the community of practitioners who constantly produce and reproduce it, in a reconciled language, in a situational territory through institutionalized practices or in institutionalization, in a paradigm of co-production and co-creation.
Experiential knowledge	Typology of knowledge generated by the tacit knowledge of experts, created on the basis of experience, resulting from undertaking social and work activities
Phenomenological knowledge	Typology of knowledge generated through the perception of phenomena by consciousness, when interpreting, from the subject's social reality, the way in which something manifests itself. This knowledge becomes knowledge about the subject themselves and their relation to the object.

Table 6.2 (*Continued*)

Term	**Description for KAD**
Strategic knowledge	Typology of organizational knowledge present in the connections and interrelationships of the elements of the system, related to planning and strategy, which seeks to align resources and capabilities for competitive advantage to be achieved. It consists not only of explicit knowledge but also of implicit and tacit knowledge of decision-makers.
General knowledge	A universally valid typology of knowledge, free from context-specific aspects, conditions and restrictions, with low dissemination cost, but whose application generally requires translation and adaptation.

Source: The authors (2021).

by three dimensions of integration (communicative, socio-organizational and cognitive-epistemic), underpinned in the search for the common good.

By completing this work, the authors, who are constructivists and co-productionists, invite the readers of this publication and users of the KAD Framework to contribute their own understandings of the meaning of transdisciplinary co-production of scientific and technological knowledge through integrative research, keeping Knowledge Acquisition Design in constant evolution and as a significant contribution. It is worth stating that this work is the result of both scientific knowledge, available in legitimate information repositories, and knowledge of the organizational field, object of master's and doctoral research studies developed at the Integration Engineering and Knowledge Governance Laboratory of the Federal University of Santa Catarina.

REFERENCES

Alford, John. 2009. *Engaging Public Sector Clients: From Service-Delivery to Co-production.* New York: Palgrave Macmillan.

———. 2014. 'The Multiple Facets of Co-production: Building on the Work of Elinor Ostrom'. *Public Management Review* 16, no. 3: 299–316.

Almeida, M. do R. G. 2000. *Literatura cinzenta: teoria e prática.* São Luís: UFMA.

Alves-Mazzotti, A. J. 2001. 'Relevância e aplicabilidade da pesquisa em educação'. *Cadernos de Pesquisa*, no. 113: 39–50.

Apostel, L., Berger, G., Briggs, A. and Michaud, G., eds. 1970. *Annales de L'Interdisciplinarité: Problèmes d'Enseignement et de Recherche dans les Universités*, Nice, 1. Paris: Organisation de Coopération et de Développement Économiques; Centre pour la Recherche et l'Innovation dans l'Enseignement, 1972.

Ariew, R. 1992. 'Descartes and the Tree of Knowledge'. *Synthese* 92, no. 1: 101–116.

Aristotle. 1989. *Metaphysics.* London: Harvard University Press.

Atten, M. Van. 2017. 'A Note on Leibniz's Argument against Infinite Wholes'. *British Journal for the History of Philosophy* 19, no. 1: 121–129.

Bacon, Francis. 1605. *The Advancement of Learning (of the Proficience and Advancement of Learning, Divine and Humane)* (VI volumes). Oxford: Leon Lieffield.

Bammer, G. 2013. *Disciplining Interdisciplinarity: Integration and Implementation Sciences for Researching Complex Real-World Problems. Transdisciplinary Perspectives on Transitions to Sustainability.* Camberra: Australian National University Press. https://doi.org/10.4324/9781315550206-13.

Bammer, G. 2019. 'What Makes a Researcher transdisciplinary? A Framework to Identify Expertise'. *Gaia Ecological Perspectives for Science and Society* 28 no. 3: 253. https://doi.org/10.14512/gaia.28.3.2

Bar-Hillel, Y. and Carnap, R. 1953. Semantic Information. *British Journal of Philosophical Sciences*, v. 4, no. 14: 147–157.

Bergmann, M., Jahn, T., Knobloch, T., Krohn, W., Pohl, C. and Schramm, E. 2012. 'Methods for Transdisciplinary Research: A Primer for practice'. In *Methods for Transdisciplinary Research: A Primer for Practice*, edited by M. Bergmann, T. Jahn, T. Knobloch, W. Krohn, C. Pohl and E. Schramm, 22–49. Frankfurt: Campus Verlag GmbH.

Bergmann, M. and Jahn, T. 'A Model for the Transdisciplinary Research Process'. *Gaia Ecological Perspectives for Science and Society* 26, no. 4: 304. https://doi.org/10.14512/gaia.26.4.3.

Bierly, P. E., Damanpour, F. and Santoro, M. D. 2009. 'The Application of External Knowledge: Organisational Conditions for Exploration and Exploitation'. *Journal of Management Studies* 46, no. 3: 482–509.

Bloor, D. 1976. *Knowledge and Social Imagery*. London: Routledge and Kegan Paul.

Boateng, H. and Agyemang, F. G. 2015. 'The Effects of Knowledge Sharing and Knowledge Application on Service Recovery Performance'. *Business Information Review* 32, no. 2: 119–126.

Bobbio, L. 2005. 'Governance multilivello e democrazia'. *Rivista delle Politiche sociali*, [S.l.], 2: 51–62.

Bobbio, N. 1997. *Os intelectuais e o poder: dúvidas e opções dos homens de cultura na sociedade contemporânea*. São Paulo: Editora da Universidade Estadual Paulista.

Bovaird, T. 2007. 'Beyond Engagement and Participation: User and Community Co-production of Public Services'. *Public Administration Review* 67, no. 5: 846–860.

Bourguinat, H. 1968. *Les marchés communs des pays en voie de développement*. Genebra: Librairie Droz.

Bourguinat, H. 1969. 'Régionalisation, Intégration, Co-production'. *Economie Appliquée* 22, no. 1–2: 245–276.

Brandsen, T. and Honingh, M. 2016. 'Distinguishing Different Types of Co-Production: A Conceptual Analysis Based on the Classical Definitions'. *Public Administration Review* 76, no. 3: 427–435. https://doi.org/10.1111/puar.12465.

Brudney, J. L. and England, R. E. 1983. 'Toward a Definition of the Co-Production Concept'. *Public Administration Review* 43, no. 1: 59. https://doi.org/10.2307/975300.

Bruhn, T., Herberg, J., Molinengo, G., Oppold, D., Stasiak, D. and Nanz, P. 2019. *Grounded Action Design: A Model of Scientific Support for Processes to Address Complex Challenges*. Potsdam: Institute for Advanced Sustainability Studies (IASS).

Brundtland, H. 1987. *The Brundtland Report: Our Common Future*. Nova York: World Commission on Environment and Development/United Nations. https://doi.org/10.1080/07488008808408783.

Bullynck, M. 2010. 'Johann Lambert's Scientific Tool Kit'. *Science in Context* 23, no. 1: 65–89.

Bunge, M. A. 1985. *Treatise on Basic Philosophy v. 7*. Dordrecht: D. Reidel Publishing Company.

Burkhardt, H. and Degen, W. 1990. 'Mereology in Leibniz's Logic and Philosophy'. *Topoi* 9, no. 1: 3–13. https://doi.org/10.1007/BF00147625.

Byé, M. 1965. 'Les formes nouvelles de la coopération internationale: vers la production en commun'. *Le Monde Diplomatique* (Decembre), 1–7.

Cabrera, D. and Cabrera, L. 2015. *Systems Thinking Made Simple: New Hope for Solving Wicked Problems*. Ithaca: Odyssean Press.

Cabrera, D. and Cabrera, L. 2018. 'Four Building Blocks of Systems Thinking'. *GAIA Ecological Perspectives for Science and Society* 27, no. 2: 200.

Cabrera, D., Cabrera, L. and Powers, E. 2015. 'A Unifying Theory of Systems Thinking with Psychosocial Applications'. *Systems Research and Behavioral Science* 32, no. 5: 534–545. https://doi.org/10.1002/sres.2351.

Campbell, J. and Pedersen, O. K. 2014. 'Knowledge Regimes and the National Origins of Policy Ideas'. In *The National Origins of Policy Ideas*, edited by J. Campbell and O. K. Pedersen, 5–26. Princeton: Princeton University Press.

Campello, B. S., Cedón, B. V. and Kremer, J. M., eds. 2000. *Fontes de informação para pesquisadores e profissionais*. Belo Horizonte: UFMG. 92–98. https://biblio-2008.webnode.com.br/_files/200000040-76a3b771d5/fontes_de_informacao_para_pesquisadores_e_profissionais_parte_001.pdf. Acesso em: 4 mar. 2021.

Cardoso, C. 2016. 'O universo monadológico: natureza, vida e expressão'. In *LEIBNIZ, G. W. Monadologia* (1a.). Lisboa: Fundação para a Ciência e a Tecnologia (FCT).

Cat, J. 2017. *Unity of Science*, edited by Stanford University. <https://plato.stanford.edu/archives/spr2022/entries/scientific-unity/>.

Checkland, P. 2000. 'Soft Systems Methodology'. *Systems Research and Behavioral Science* 17, S11–S58.

Chen, C.-J. and Huang, J.-W. 2009. 'Strategic Human Resource Practices and Innovation Performance: The Mediating Role of Knowledge Management Capacity'. *Journal of Business Research* 62, no. 1: 104–114.

Chesbrough, H. W. 2003. *Open Innovation: The New Imperative for Creating and Profiting from Technology*. Boston: Harvard Business Press.

Chomsky, N. 2014. *What Is the Common Good?* Nova York: Columbia University.

Clayton, E. 1995. 'Aristotle: Politics'. In *The Internet Encyclopedia of Philosophy* (1st ed., p. s/p). The Internet Encyclopedia of Philosophy. Retrieved from s/u. ISSN 2161-0002.

Cochrane Effective Practice and Organisation of Care (Epoc). 2020. *EPOC Qualitative Evidence Syntheses: Protocol and Review Template*. London: Cochrane. https://epoc.cochrane.org/sites/epoc.cochrane.org/files/public/uploads/Resources-for-authors2017/epoc_qes_protocol_and_review_template.docx.

Collins, H. M. 1985. *Changing Order: Replication and Induction in Scientific Practice*. Chicago: University of Chicago Press.

Cornwall, A. and Jewkes, R. 1995. 'What Is Participatory Research?' *Social Science & Medicine* 41, no. 12: 1667–1676. https://doi.org/10.2167/md073.0.

Couto, R. M. 2018. *Governança nas instituições de ensino superior: análise dos mecanismos de governança na Universidade Federal de Santa Catarina à luz do modelo multilevel governance* (Dissertação de Mestrado). Universidade Federal de Santa Catarina.

Creswell, J. W. 2007. *Projeto de pesquisa: métodos qualitativo, quantitativo e misto* (2nd ed.). Porto Alegre: Artmed/Bookman.

Crossan, M. M., Lane, H. W. and White, R. E. 1999. 'An Organisational Learning Framework: From Intuition to Institution'. *Academy of Management Review* 24, no. 3: 522–537.

Cupani, A. 2006. 'La peculiaridad del conocimiento tecnológico'. *Scientiae Studia* 4, no. 3: 353–371. https://doi.org/10.1590/s1678-31662006000300002.

Descartes, R. 1644. *Principia philosophiae*. Amstelodami: Ludovicum Elzevirium.

———. 1998. *Regula ad directionem ingenii (Rules for the direction of the natural intelligence): a bilingual edition of the cartesian treatise on method*. Amsterdam: Rodopi.

Descartes, R. and Maclean, I. 2006. *A Discourse on the Method of Correctly Conducting One's Reason and Seeking Truth in the Sciences. Oxford World's Classics*. Oxford: Oxford University Press.

Diderot, D. and D'Alembert, Jean Le R. 1751–1772. *Encyclopédie ou dictionnaire raisonné des sciences, des arts et des métiers: par une societé de gens de lettres*. Paris: André Le Breton, Laurent Durand, Antoine-Claude Briasson, Michel-Antoine David.

Enengel, B., Muhar, A., Penker, M., Freyer, B., Drlik, S. and Ritter, F. 2012. 'Co-Production of Knowledge in Transdisciplinary Doctoral Theses on Landscape Development: An Analysis of Actor Roles and Knowledge Types in Different Research Phases'. *Landscape and Urban Planning* 105, no. 1–2: 106–117.

Elkana, Y. 1979. 'Science as a Cultural System: An Anthropological Approach. In *Scientific Culture in the Contemporary World*, edited by N. Bonetti, 269–290. Milano: Scientia.

Engeström, Y. 1987. *Learning by Expanding: An Activity-Theoretical Approach to Developmental Research.* Helsinki: Orienta-Konsultit.

Fazey, I., Schäpke, N., Caniglia, G., Patterson, J., Hultman, J., van Mierlo, B. and Wyborn, C. 2018. 'Ten Essentials for Action-Oriented and Second Order Energy Transitions, Transformations and Climate Change Research'. *Energy Research and Social Science* 40, 54–70. https://doi.org/10.1016/j.erss.2017.11.026.

Fennell, L. A. 2011. 'Ostrom's Law: Property Rights in the Commons Lee'. *International Journal of the Commons* 5, no. 1: 9–27. https://doi.org/10.1080/00222895.1991.9941592

Fidel, R. 1984. 'The Case Study Method: A Case Study'. *Library and Information Science Research* 6, no. 3: 273–288.

Fleury, M. T. and Oliveira Jr., M. M., eds. 2001. *Gestão estratégica do conhecimento: integrando aprendizagem, conhecimento e competências.* São Paulo: Atlas.

Floridi, L. 2015. 'Semantic Conceptions of Information'. In *Stanford Encyclopedia of Philosophy.* Center for the Study of Language and Information (CSLI).

Foss, N. J. and Mahoney, J. T. 2010. 'Exploring Knowledge Governance'. *International Journal Strategic Change Management* 2, no. 2/3: 93–101.

Freire, P. de S., Alvares, L. M. A. de R., Rizzatti, G., Bresolin, G. G., Martins, G. T., Silva, T. C. and Kempner-Moreira, F. 2021. *Glossário:governança multinível do conhecimento e da aprendizagem e seus mecanismos de universidade corporativa em rede e de comunicação dialógica* (1st ed.). Curitiba: CRV Editora.

Freire, P. de S., Alvares, L. M. A. de R. and Silva, S. M. da. 2021. 'Conhecimentos críticos para a prontidão tecnológica da inovação: a inter-relação das informações tecnológicas e do ciclo tecnológico'. In *Os múltiplos cenários da informação tecnológica no Brasil do século XXI,* edited by L. M. A. de R. Alvares and A. L. C. Itaborahy, 429–469. Brasília: IBICT: UNESCO.

Freire, P. de S., Kempner-Moreira, F. and Hott Jr., J. L. 2020. 'Governança multinível em rede: reflexões sobre um novo modelo de governança para a segurança pública'. In *Anais do VII Encontro Brasileiro de Administração Pública,* 1–16. Brasília, Brasil: EBAP.

Freire, P. de S. et al. 2017. 'Governança do conhecimento (GovC): o estado da arte sobre o termo'. *Biblios* 69, 21–40.

Frodeman, R. 2011. 'Interdisciplinary Research and Academic Sustainability: Managing Knowledge in an Age of Accountability'. *Environmental Conservation* 38, no. 2: 105–112. https://doi.org/10.1017/S0376892911000038.

———. 2014. *Sustainable Knowledge: A Theory of Interdisciplinarity. Sustainable Knowledge: A Theory of Interdisciplinarity.* Hampshire: Palgrave Macmillan. https://doi.org/10.1057/9781137303028.

Garcé, A. 2015. 'Political-Knowledge Regimes: Building a New Concept from Selected Policy-Change Events in the Tabaré Vázquez Administration (Uruguay, 2005–2009)'. *World Political Science Review* 11, no. 1: 23–45. https://doi.org/10.1515/wpsr-2014-0020.

Ghassib, H. 2012. 'A Theory of the Knowledge Industry'. *International Studies in the Philosophy of Science* 26, no. 4: 447–456. https://doi.org/10.1080/02698595.2012.748499.

Gherardi, S. 2008. 'Situated Knowledge and Situated Action: What do Practice-Based Studies Promise?' *The SAGE Handbook of New Approaches in Management and organisation,* (March), 516–525. https://doi.org/10.4135/9781849200394.n89.

Gibbons, M., Limoges, C., Nowotny, H., Schwartzman, S., Scott, P. and Trow, M. 1994. *The New Production of Knowledge: The Dynamics of Science and Research in Contemporary Societies.* London: Sage.

Gil, A. C. 2009. *Como elaborar projetos de pesquisa* (4th ed.). São Paulo: Atlas.

Groß, V. and Hoffmann-Riem, V. 2005. 'Ecological Restoration as a Real-World Experiment: Designing Robust Implementation Strategies in an Urban Environment'. *Public Understanding of Science*, 14, no. 3: 269–284.

Godemann, J. 2008. 'Knowledge Integration: A Key Challenge for Transdisciplinary Cooperation'. *Environmental Education Research* 14, no. 6: 625–641. https://doi.org/10.1080/13504620802469188.

Goffman, E. 1971. 'The Territories of the Self'. In *Relations in Public: Microstudies of the Public Order*, edited by E. Goffman, New York: Harper & Row.

Hadorn, G. H., Biber-Klemm, S., Grossenbacher-Mansuy, W., Hoffmann-Riem, H., Joye, D., Pohl, C., Wiesmann, U. and Zemp, E. 2008. *Handbook of Transdisciplinary Research*. Berlim: Springer.

Hadorn, G. H., Pohl, C. and Bammer, G. 2010. 'Solving Problems Through Transdisciplinary Research'. In *The Oxford Handbook of Interdisciplinarity*, edited by R. Frodeman, J. T. Klein and C. Mitcham, 431–452. Oxford: Oxford University Press.

Hansson, S. O. 2013. 'What Is Technological Knowledge'. In *Technology Teachers as Researchers*, edited by I.-B. Skogh and M. J. De Vries, 305. Rotterdam: Sense Publishers.

Hegel, Georg W. F. 1812. *Wissenschaft der logik*. (Bd.1,2 – 1813; Bd 2 – 1816). Nürnberg.

Heisenberg, W. 1989. *Ordnung der wirklichkeit*. Munich: Piper Verlag.

Held, V. 1970. *The Public Interest and Individual Interests*. New York: Basic Books.

Hevner, A. R. 2007. 'A Three Cycle View of Design Science Research'. *Scandinavian Journal of Information Systems* 19, no. 2: 1–6.

Higgins, J. and Thomas, J., eds. 2019. *Cochrane Handbook for Systematic Reviews of Interventions*. Hoboken: John Wiley & Sons.

Hirata, C. 2012. 'Sistema em Leibniz e Descartes'. *Trans/Form/Ação 35*, no. 1: 23–36. https://doi.org/10.1590/S0101-31732012000100003.

Huber, G. P. 1991. 'Organizational Learning: The Contributing Processes and the Literatures'. *Organization Science*, 2, no. 1: 88–115.

Hunter, L. 2009. 'Situated Knowledge'. In *Mapping Landscapes for Performance as Research* (1st ed.), edited by S. R. Riley and L. Hunter, Hampshire (Inglaterra): Palgrave Macmillan. https://doi.org/10.1057/9780230244481.

Hussain, W. 2018. *The Common Good*, edited by Stanford University. https://plato.stanford.edu/archives/spr2018/entries/common-good/.

Indiana University Bloomington. (n.d.). *Ostrom Workshop: History*. Bloomington: IUB.

Jaede, M. 2017. *The Concept of the Commons* (1). *Series of the Political Settlements Research Programme (PSRP)*. Edimburgo: University of Edinburgh. https://doi.org/10.2307/j.ctt183q67g.7.

Jahn, T. 2005. 'Soziale ökologie, kognitive integration und transdisziplinarität'. *TATuP: Zeitschrift für Technikfolgenabschätzung in Theorie und Praxis* 14, no. 2: 32–38.

———. 2008. 'Transdisciplinarity in the Practice of Research'. In *Transdisziplinäre Forschung. Integrative forschungsprozesse verstehen und bewerten*, edited by M. Bergmann and E. Schramm, Vol. 37, 21–37. Frankfurt: Campus Verlag GmbH.

Jantsch, E. 1972. Vers l'interdisciplinarité et la transdisciplinarité dans l'enseignement et l'innovation. In Annales de l'*Interdisciplinarité: Problèmes d'Enseignement et de Recherche dans les Universités. 1970*. Paris: Organisation de Coopération et de Développement Économiques; Centre pour la Recherche et l'Innovation dans l'Enseignement.

Jasanoff, S. 2004a. 'Ordering Knowledge, Ordering Society'. In *States of Knowledge: The Co-Production of Science and Social Order*, edited by S. Jasanoff. London: Routledge.

———. 2004b. *States of Knowledge: The Co-Production of Science and Social Order*, edited by S. Jasanoff. London: Routledge.

———. 2004c. 'The Idiom of Co-Production'. In *States of Knowledge: The Co-Production of Science and Social Order*, edited by S. Jasanoff. London: Routledge.

Kempner-Moreira, F. and Freire, P. de S. 2020. 'The Five Stages of Evolution of Interorganisational Networks: A Review of the Literature'. *Journal of Information & Knowledge Management* 19, no. 4: 2050038-1-2050038-19.

Kiernan, M. J. 1998. *Os 11 mandamentos da administração do século XXI*. São Paulo: Makron Books.

Kiser, L. and Percy, S. L. 1980. 'The Concept of Co-Production and Its Prospects for Public Service Delivery'. In *Workshop in Political Theory and Policy Analysis*. San Francisco: Annual Meetings of the American Society for Public Administration.

Klein, J. 2013. 'The Transdisciplinary Moment(um)'. *Integration Review* 9, no. 2: 189–199.

Klein, J. T. 2009. 'Unity of Knowledge and Transdisciplinarity: Contexts of Definition, Theory and the New Discourse of Problem Solving'. In *Unity of Knowledge (in Transdisciplinary Research for Sustainability)*, vol. I. Paris: UNESCO (EOLSS Encyclopedia of Life Support Systems Publications).

———. 2010. 'A Taxonomy of Interdisciplinarity'. In *The Oxford Handbook of Interdisciplinarity*, edited by R. Frodeman, J. T. Klein and C. Mitcham (2nd ed.), 15–30. Oxford: Oxford University Press Inc.

Knowles, M. S. 1980. *The Modern Practice of Adult Education: Andragogy Versus Pedagogy*. Englewood Cliffs: Prentice Hall.

Kockelmans, J. J. 1979. 'Why Interdisciplinarity?' In *Interdisciplinarity and Higher Education*, edited by J. J. Kockelmans. Pennsylvania: Penn State University Press.

Kolb, D. A. 1984. *Experiential Learning: Experience as the Source of Learning and Development*. New Jersey: Prentice Hall.

Kötter, R. and Balsiger, P. W. 1999. 'Interdisciplinarity and Transdisciplinarity: A Constant Challenge to the Sciences'. *Issues in Integrative Studies* 17, 87–120.

Kroes, P. 1989. 'Philosophy of Science and the Technological Dimension of Science'. In *Imre Lakatos and Theories of Scientific Change*, edited by K. Gavroglu, Y. Goudaroulis and P. Nicolacopoulos, 375–381. Dordrecht: Kluwer Academic Publishers.

Kuhn, Thomas S. 2012. *The Structure of Scientific Revolutions*. Chicago: University of Chicago Press.

Lakatos, E. M. and Marconi, M. de A. 1991. *Metodologia científica* (2nd ed.). São Paulo: Atlas.

Lambert, J. H. *Neues organon oder gedanken über die erforschung und bezeichnung des wahren und dessen unterscheidung vom irrtum und schein*. Leipzig: Johann Wendler, 1764.

Latour, B. 1987. *Science in Action: How to Follow Scientists and Engineers Through Society*. Cambridge (Estados Unidos): Harvard University Press.

———. 1993. *We Have Never Been Modern*. Cambridge: Harvard University Press. https://doi.org/10.1016/0956-5221(96)88504-6.

Latour, B. and Woolgar, S. 1979. *Laboratory Life: The Social Construction of Scientific Facts*. Los Angeles: Sage.

Lawrence, R. J. 2004. 'Housing and Health: From Interdisciplinary Principles to Transdisciplinary Research and Practice. *Futures* 36, 487–502.

Leibnüzio, G. G. 1666. *Dissertatio de arte combinatoria*. Lipsiae: Joh.Simon Fickium et Joh. Polycarp. Seuboldum.

Leibniz, G. W. 2008. *New Essays on Human Understanding*. Oxford: Early Modern Texts.

Lewin, K. 1946. 'Action Research and Minority Problems'. *Journal of Social Issues* 2, no. 4: 34–46. https://doi.org/10.1111/j.1540-4560.1946.tb02295.x.

Liao, S. H., Chang, W. J., Hu, D. C. and Yueh, Y. L. 2012. 'Relationships among Organisational Culture, Knowledge Acquisition, Organisational Learning, and Organisational Innovation in Taiwan's Banking and Insurance Industries'. *International Journal of Human Resource Management* 23, no. 1: 52–70. https://doi.org/10.1080/09585192.2011.599947.

Lichnerowicz, A. 1980. Mathématiques et transdisciplinarité. *Études Renaniennes, Colloque D*, no. 43, 22–32.

Luca, Felipe A. de. 2016. 'Apontamentos para uma epistemologia leibniziana'. *Kalagatos*, Fortaleza, 55–68, v.13, n. 26. ISSN: 1984-9206.

Lulli, Raymundi (Illuminati Patris, Maioricensis). 1515. *Arbor scientiae: venerabilis et caelitus* (2a. ed.). Lyon: Guilhelmi Huyon & Constantini Fradin (Llull, Ramon. *Arbre de la ciència*. (1a. ed.). Rome: 1295–1296).

Lusthaus, C., Anderson, G. and Murphy, E. 1995. *Institutional Assessment: A Framework for Strengthening Organisational Capacity for IDRC's Research Partners*. Ottawa: IDRC.

Machlup, F. 1962. *The Production and Distribution of Knowledge in the United States*. Princeton: Princeton University Press.

Marconi, M. de A. and Lakatos, E. M. 2009. *Metodologia do trabalho científico: procedimentos básicos de pesquisa bibliográfica, projeto e relatório* (7th ed.). São Paulo: Atlas.

Marques, M. da. C. da C. 2007. 'Aplicação dos princípios da governança corporativa ao setor público'. *Revista de Administração Contemporânea (RAC)* 11, no. 2: 11–26. https://doi.org/10.1590/S1415-65552007000200002.

Martin, S. G. 1926. 'History of Philosophy'. *Monist* 36, no. 4: 678–699. https://doi.org/10.5840/monist192636435.

Matthews, M. R. (Ed.). 1989. *The Scientific Background to Modern Philosophy: Selected Readings*. Indianapolis: Hackett Publishing.

Maxwell, J. A. 2012. *A Realist Approach for Qualitative Research*. Thousand Oaks, CA: Sage Publications.

———. 2013. *Qualitative Research Design: An Interactive Approach*. Thousand Oaks, CA: Sage Publications.

McCain, K. and Kampourakis, K. 2020. *What Is Scientific Knowledge? An Introduction to Contemporary Epistemology of Science*. New York: Routledge. https://doi.org/10.1086/286587.

McMeekin, N., Wu, O., Germeni, E. and Briggs, A. 2020. 'How Methodological Frameworks are Being Developed: Evidence from a Scoping Review'. *BMC Medical Research Methodology* 20, no. 1: 1–9. https://doi.org/10.1186/s12874-020-01061-4.

Mendelsohn, E. 1977. 'The Social Construction of Scientific Knowledge'. In *The Social Production of Scientific Knowledge* (Vol. 1), edited by E. Mendelsohn, P. Weingart and R. Whitley. Dordrecht: D. Reidel Publishing Company.

Merton, Robert K. 1938. 'Science, Technology and Society in Seventeenth Century England. *Osiris* 4, 360–632.

Midgely, G. 2003. 'Science as Systemic Intervention: Some Implications of Systems Thinking and Complexity for the Philosophy of Science'. *Cuadernos de Administración* 16, no. 25: 7–30.

Miller, R. C. 1982. 'Varieties of Interdisciplinary Approaches in the Social Sciences: A 1981 Overview'. *Issues in Integrative Studies* 37, no. 1: 1–37.

Mitchell, C., Cordell, D. and Fam, D. 2015. 'Beginning at the End: The Outcome Spaces Framework to Guide Purposive Transdisciplinary Research'. *Futures* 65, 86–96. https://doi.org/10.1016/j.futures.2014.10.007.

Mobjörk, M. 2010. 'Consulting versus Participatory Transdisciplinarity: A Refined Classification of Transdisciplinary Research. *Futures* 42, no. 8: 866–873.

Morgan, G. 1996. *Imagens da organização* (1. ed.). São Paulo: Atlas.

Morin, E. 1990. *Introduction à la pensée complexe*. Paris: ESF éditeur.

Morin, E. 2005. *Introduction à la pensée complexe. Points 534* (2a.). Paris: Éditions du Seuil.

National Grey Literature Collection. 1997. *Luxembourg Definition*. London: Health Education England. http://allcatsrgrey.org.uk/wp/knowledgebase/luxembourg-definition-1997/

———. 2010. *Prague Definition*. London: Health Education England. http://allcatsrgrey.org.uk/wp/knowledgebase/prague-definition-2010/.

Nicolescu, B. 2006. *Transdisciplinarity: Past, Present and Future*. In *Moving Worldviews: Reshaping Sciences, Policies and Practices for Endogenous Sustainable Development*, edited by B. Haverkort and C. Reijntjes. Holanda: Compas Editions.

———. 1996a. 'Godelian Aspects of Nature and Knowledge'. In *Systems: New Paradigms for the Human Sciences*, edited by G. Altmann and W. A. Koch, 385–403. Paris: Universite Pierre et Marie Curie. Division de Physique Theorique. https://doi.org/10.1515/9783110801194.385.

———. 1996b. *La transdisciplinarité: manifeste* (1a. ed.). Monaco: Éditions du Rocher.

Nonaka, I. and Takeuchi, H. 2004. *Criação de conhecimento na empresa: como as empresas japonesas geram a dinâmica da inovação* (18a. ed.). Rio de Janeiro: Campus.

Norman, J. M. 2020. 'History of Information: Exploring the History of Information and Media through Timelines'. In *Llull's Tree of Knowledge* (Dictionary of Scientific Biography VIII (1973), 547). Novato, CA: Jeremy Norman & Co., Inc.

Nowotny, H., Scott, P. B. and Gibbons, M. T. 2013. 'From Reliable Knowledge to Socially Robust Knowledge'. In *Re-thinking Science: Knowledge and the Public in an Age of Uncertainty* (1a ed.), edited by H. Nowotny, P. B. Scott and M. T. Gibbons. Hoboken: John Wiley & Sons.

Ogawa, R. T. and Malen, B. 1991. 'Towards Rigor in Reviews of Multivocal Literatures: Applying the Exploratory Case Study Method'. *Review of Educational Research* 61, no. 3: 265–286.

Ollaik, L. G. and Ziller, H. M. 2012. 'Concepções de validade em pesquisas qualitativas'. *Educação e Pesquisa* 38, no. 1: 229–241.

Ostrom, E. 1990. *Governing the commons: the evolution of institutions for collective action*. Cambridge: Cambridge University Press.

———. 1996. 'Crossing the Great Divide: Co-production, Synergy, and Development'. *World Development* 24, no. 6: 1073–1087.

Pacheco, R., Santos, N. and Wahrhaftig, R. 2020. 'Transformação digital na educação superior: modos e impactos na universidade'. *Revista Nupem* 12, no. 27: 94–128. https://doi.org/10.33871/nupem.2020.12.27.94-128.

Papa, A., Dezi, L., Gregori, G. L., Mueller, J. and Miglietta, N. 2018. 'Improving Innovation Performance through Knowledge Acquisition: The Moderating Role of Employee Retention and Human Resource Management Practices'. *Journal of Knowledge Management* 24, no. 3: 589–605. https://doi.org/10.1108/JKM-09-2017-0391.

Parks, R. B., Baker, P. C., Kiser, L., Oakerson, R., Ostrom, E., Ostrom, V. and Wilson, R. 1981. 'Consumers as Co-Producers of Public Services: Some Economic and Institutional Considerations'. *Policy Studies Journal* 9, no. 7: 1001–1011.

Patton, M. Q. 1988. 'Paradigms and Pragmatism'. In *Qualitative Approaches to Evaluation in Education: The Silent Scientifique Revolution Analysis*, edited by D. M. Fetterman. Nova York: Praeger.

Pemsel, S. and Müller, R. 2012. 'The Governance of Knowledge in Project-Based Organisations'. *International Journal of Project Management* 30, no. 8: 865–876.

Percy, S. L. 1978. 'Conceptualizing and Measuring Citizen Co-Production of Community Safety'. *Policy Studies Journal* 7, no. 51: 486–493.

Pereira, M. F. 2010. *Planejamento estratégico: teorias, modelos e processos.* São Paulo: Atlas.

Pestoff, Vi. 2006). Citizens as co-producers of welfare services: preschool services in eight European countries. *Public Management Review*, 8(4), 503–20.

———. 2009. *A Democratic Architecture for the Welfare State.* London: Routledge.

Piaget, J. 1972. 'L'épistémologie des relations interdisciplinaires'. In *L'Interdisciplinarité: Problèmes d'Enseignement et de Recherche dans les Universités. 1970.* Paris: Organisation de Coopération et de développement économiques; Centre pour la recherche et l'innovation dans l'enseignement.

Piattoni, S. 2009. 'Multi-Level Governance: A Historical and Conceptual Analysis'. *Journal of European Integration* 31, no. 2: 163–180.

Población, D. A. 1992. 'Literatura cinzenta ou não convencional: um desafio a ser enfrentado'. *Ciência da Informação* 21, no. 3: 243–246. http://revista.ibict.br/ciinf/article/view/438. Acesso em: 4 mar. 2021.

Pohl, C. 2010. 'From Transdisciplinarity to Transdisciplinary Research'. *Transdisciplinary Journal of Engineering and Science (TJES)* 1, no. 1: 65–73. https://doi.org/10.22545/2010/0006

Pohl, C. and Hadorn, G. H. 2017. 'Principles for Designing Transdisciplinary Research'. *GAIA Ecological Perspectives for Science and Society* 26, no. 3: 232–232.

Pohl, C., Kerkhoff, L. van, Hadorn, G. H. and Bammer, G. 2018. 'Integration'. In *Handbook of Transdisciplinary Research*, edited by G. H. Hadorn, S. Biber-Klemm, W. Grossenbacher-Mansuy, H. Hoffmann-Riem, D. Joye, C. Pohl, U. Wiesmann and E. Zemp, 411–426. Berlin: Springer.

Pohl, C., Rist, S., Zimmermann, A., Fry, P., Gurung, G. S., Schneider, F. and Urs, W. 2010. 'Researchers' Roles in Knowledge Co-Production: Experience from Sustainability Research in Kenya, Switzerland, Bolivia and Nepal'. *Science and Public Policy* 37, no. 4: 267–281. https://doi.org/10.3152/030234210X496628

Polk, M. 2011. 'Institutional Capacity-Building in Urban Planning and Policy-Making for Sustainable Development: Success or Failure?' *Planning, Practice & Research* 26, no. 2: 185–206.

Polk, M. 2015. 'Transdisciplinary Co-Production: Designing and Testing a Transdisciplinary Research Framework for Societal Problem Solving'. *Futures* 65, 110–122. https://doi.org/10.1016/j.futures.2014.11.001

Preece, J., Rogers, Y. and Sharp, H. 2005. *Design de interação: além da interação humano-computador* (1st ed.). Porto Alegre: Bookman.

Putnam, R. 1993. 'The Prosperous Community: Social Capital and Public Life'. *The American Prospect* 13, 35–42.

Ramadier, T. 2004. 'Transdisciplinarity and Its Challenges: The Case of Urban Studies'. *Futures* 36, 423–439.

Ramaswamy, V. and Ozcan, K. 2014. *The Co-Creation Paradigm.* Stanford: Stanford University Press.

Ravitch, S. M. and Riggan, M. 2017. *Reason & Rigor: How Conceptual Frameworks Guide Research.* Thousand Oaks, CA: Sage Publications.

Ravitch, Sharon M. and Carl, N. M. 2016a. 'Conceptual Frameworks in Research'. In *Qualitative Research: Bridging the Conceptual, Theoretical and Methodological*, 32–61. Thousand Oaks, CA: Sage Publications.

Ravitch, Sharon M and Carl, N. M. (2016b). *Qualitative Research: Bridging the Conceptual, Theoretical, and Methodological*. Thousand Oaks, CA: Sage Publications.

Rizzatti, G. 2020. *Framework de governança da aprendizagem organizacional*. 2020. (Tese de Doutorado). Universidade Federal de Santa Catarina, Florianópolis, SC.

Rizzatti, G. and Freire, P. de S. 2020. 'Governança da aprendizagem organizacional (GovA): o estado da arte sobre o termo'. *Espacios* 41, no. 3: 16.

Roesch, S. M. A. 2009. *Projetos de estágio e de pesquisa em administração: guia para estágios, trabalhos de conclusão, dissertações e estudos de caso* (3rd ed.). São Paulo: Atlas.

Rosenfield, P. L. 1992. 'The Potential of Transdisciplinary Research for Sustaining and Extending Linkages between the Health and Social Sciences'. *Social Science & Medicine* 35, 1343–1357.

Santos, A. 2008. 'Complexidade e transdisciplinaridade em educação'. *Revista Brasileira de Educação* 13, no. 37: 71–84.

Scholz, R. W. and Tietje, O. 2002. *Embedded Case Study Methods*. London: Sage Publications.

Senge, P. M. 1990. *The Fifth Discipline: The Art and Practice of the Learning Organisation*. London: Random House Business.

Shannon, C. E. 1993. *Collected Papers*, edited by N. J. A. Sloane and A. D. Wyner. New York: IEEE Press.

Sharp, E. B. 1980. 'Toward a New Understanding of Urban Services and Citizen Participation: The Co-Production Concept'. *Midwest Review of Public Administration* 14, no. 2: 105–118.

Shujahat, M., Sousa, M. J., Hussain, S., Nawaz, F., Wang, M. and Umer, M. 2019. 'Translating the Impact of Knowledge Management Processes into Knowledge-Based Innovation: The Neglected and Mediating Role of Knowledge-Worker Productivity'. *Journal of Business Research* 94: 442–450.

Simm, K. 2011. 'The Concepts of Common Good and Public Interest: From Plato to Biobanking'. *Cambridge Quarterly of Healthcare Ethics* 20, no. 4: 554–562. https://doi.org/10.1017/S0963180111000296.

Smith, R. 2017. *Aristotle's Logic*, edited by Stanford University. https://plato.stanford.edu/archives/fall2020/entries/aristotle-logic/

Sonnenwald, D. H. 2007. 'Scientific Collaboration'. *Annual Review of Information Science and Technology* 41, no. 1: 643–681. https://doi.org/10.1002/aris.2007.1440410121.

Sperandio, D. J. and Évora, Y. D. M. 2005. 'Planejamento da assistência de enfermagem: proposta de um software-protótipo'. *Revista Latino-Americana de Enfermagem Enfermagem* 13, no. 6: 937–943.

Stokols, D., Hall, K. L., Moser, R. P., Feng, A., Misra, S. and Taylor, B. K. 2010. 'Cross-Disciplinary Team Science Initiatives: Research, Training, and Translation'. In *The Oxford Handbook of Interdisciplinarity* (2nd ed.), edited by R. Frodeman, J. T. Klein and C. Mitcham. Oxford: Oxford University Press.

Taylor, Edward W. 1998. *The Theory and Practice of Transformative Learning: A Critical Review*. Columbus: ERIC Clearinghouse on Adult, Career, and Vocational Education, Center on Education and Training for Employment, College of Education, the Ohio State University.

The Nobel Prize. 2009. *Elinor Ostrom Facts* (The Sveriges Riksbank Prize in Economic Sciences in Memory of Alfred Nobel 2009).

Tress, G., Tress, B. and Fry, G. 2005. 'Clarifying Integrative Research Concepts in Landscape Ecology'. *Landscape Ecology* 20, no. 4: 479–493. https://doi.org/10.1007/s10980-004-3290-4.

Turner, J. 1997. *The Institutional Order*. New York: Longman.

Umpleby, S. 2014. 'Second-Order Science: Logic, Strategies, Methods'. *Constructivist Foundations (Special Issue on Second Order Science)* 10, no. 1: 16–23.

Varzi, A. 2016. *Mereology*. In *Stanford Encyclopedia of Philosophy*, edited by Stanford University. https://plato.stanford.edu/archives/spr2019/entries/mereology/.

Vaegs, T., Zimmer, I., Schröder, S., Leisten, I., Vossen, R. and Jeschke, S. 2013. 'Fostering Interdisciplinary Integration in Engineering Management'. *Proceedings of the IEEE International Conference on Industrial Engineering and Engineering Management*, (December 2015), 23–28. https://doi.org/10.1109/IEEM.2013.6962367.

Vera, D. and Crossan, M. 2004. 'Strategic Leadership and Organisational Learning'. *Academy of Management Review* 29, no. 2: 222–240.

Vergara, S. C. 2005. *Projetos e relatórios de pesquisa em administração* (4th ed.). São Paulo: Atlas.

Weingart, P. 2010. 'A Short History of Knowledge Formations'. In *The Oxford Handbook of Interdisciplinarity* 2a., edited by R. Frodeman, J. T. Klein and C. Mitcham, 3–14. Oxford: Oxford University Press Inc.. https://doi.org/10.31046/tl.v5i2.228.

West, J. and Bogers, M. 2014. 'Leveraging External Sources of Innovation: A Review of Research on Open Innovation'. *Journal of Product Innovation Management* 31, no. 4: 814–831.

Weyrauch, V., Echt, L. and Suliman, S. 2016. *Knowledge into Policy: Going Beyond 'Context Matters'* (1st ed.). Dublin: Purpose & Ideas (P&I).

Wiener, N. 1954. *The Human Use of Human Beings: Cybernetics and Society*, 15–27. Boston: Houghton Mifflin.

Whitaker, G. P. 1980. 'Co-Production: Citizen Participation in Service Delivery'. *Public Administration Review* 40, no. 3: 240–246.

Wray, K. B. 2002. 'The Epistemic Significance of Collaborative Research'. *Philosophy of Science* 69, no. 1: 150–168. https://doi.org/10.1007/s11098-015-0537-7.

Zahra, S. A. and George, G. 2002. 'Absorptive Capacity: A Review, Reconceptualization, and Extension'. *Academy of Management Review* 27, no. 2: 185–203.

Zins, C. 2007. 'Conceptual Approaches for Defining Data, Information, and Knowledge'. *Journal of the American Society for Information Science and Technology* 58, no. 4: 479–493. https://doi.org/10.1002/asi.

GLOSSARY

COMMON GOOD: shared social vision, with the objective of defining a way of thinking and acting that reflects a mutual concern among the members of a community, with a view to promoting the development of humans, in their richest diversity, advancing their rights, well-being and fulfilment of fair aspirations, providing them with a more dignified, responsible and productive way of life (Freire et al. 2021, p. 43).

COMPLEXITY: an approach that avoids fragmentation, through systemic reasoning, by integrating the antagonistic poles of the contradiction of reality, multiplicity, randomness and uncertainty; it is better explained by three underlying principles: the Principle of organizational Recursion, the Dialogic Principle and the Holographic Principle (Nicolescu 1996b).

SCIENTIFIC KNOWLEDGE: typology of knowledge considered to be true by science, produced in a transparent and traceable way, based on solid foundations, acquired through scientific research, meeting strict epistemic standards, whose results are reliable, robust and well-established on a certain subject at a given point in time.

STRATEGIC KNOWLEDGE: typology of organizational knowledge present in the connections and interrelations of the elements of the system, relative to planning and strategy, which seeks to align resources and capabilities to achieve competitive advantage, consisting of both explicit knowledge and implicit and tacit knowledge of decision-makers.

EXPERIENTIAL KNOWLEDGE: typology of knowledge generated by the tacit knowledge of specialists, created on the basis of experience, resulting from doing social and work activities.

PHENOMENOLOGICAL KNOWLEDGE: typology of knowledge generated through the conscious perception of phenomena, when one interprets, based on the subject's social reality, the way in which something manifests itself; thus, such knowledge becomes knowledge about the subject themselves and their relation with the object.

SITUATED KNOWLEDGE: typology of knowledge located in an organizational structure, present in the relations established in the community of practitioners who constantly produce and reproduce it, in a reconciled language, in a situational territory through practices that are already institutionalized or are being institutionalized, in a paradigm of co-production and co-creation.

SUSTAINABLE KNOWLEDGE: the perception that the production of knowledge has become unsustainable owing to the infinite production of scientific knowledge, because most researchers (and people in general) do not need as much information as is currently available. The term sustainable knowledge is sometimes used as an alternative to transdisciplinary knowledge (Frodeman 2014).

CO-PRODUCTION: the process through which the inputs used to produce a good or a service are provided by individuals who are not in the same organization (Ostrom 1996).

TRANSDISCIPLINARY CO-PRODUCTION: taking place in the interaction between the scientific and social perspectives, it supersedes the disciplinary divisions of science and technology by creating a fertile field of dialogue, in which the multiple realities of science and social practices are perceived, involving both the scientific method and the context where the problem occurs.

KNOWLEDGE ACQUISITION DESIGN: framework for designing and carrying out integrative research on transdisciplinary co-production in knowledge governance and organizational learning, which is structured through processes of action and transformation contained in three moments of integration (situation, creation and solution), based on the communicative, socio-organizational and cognitive-epistemic dimensions, whose guiding principle is the pursuit of the common good.

MODE 2: knowledge intended to be useful to someone, whether in industry or government, or in society at large, always produced under an aspect of continuous negotiation; it will not be produced unless and until the interests of the various actors have been included (Gibbons et al. 1994).

INTEGRATIVE RESEARCH: interdisciplinary and transdisciplinary research, whose particular feature is the recognition that there are varying degrees of stakeholders' integration and involvement in integrative approaches (Bergmann et al. 2012).

TRANSDISCIPLINARY RESEARCH: integrative scientific research which unites academic knowledge and situated knowledge in order for individuals to gain a perception of essential traits that point to the problem. The communicative, socio-organizational and cognitive-epistemic dimensions are needed for integration to occur, in a permanent pursuit of the unity of knowledge, with the objective of supporting the transformations that are called for in society.

PRINCIPLE OF organizational RECURSION: refers to processes that are both products and producers, that is, everything that is produced returns to what produces it in a recursive cycle contrary to the idea of linearity (Morin 1990).

DIALOGIC PRINCIPLE: refers to the two logics that are at the same time complementary and antagonistic, acting in the same system, collaborating and producing organization and complexity, allowing the duality to be maintained within the unit (Morin 1990).

HOLOGRAPHIC PRINCIPLE: refers to the relationship between the whole and its parts, especially to the fact that not only the part is in the whole, but the whole is in the part (Morin 1990).

TRANSDISCIPLINARITY: the pursuit of the unity of knowledge across disciplinary boundaries, in order to capture the full complexity of the multidimensional and the multireferential Reality of the conditioned element, by addressing relevant social issues, providing deep levels of transformation in higher education with a view to co-production of scientific knowledge, aiming at the common good.

UNITY OF KNOWLEDGE: consilient knowledge, whose unity can only be achieved through conditioned knowledge and unconditioned knowledge, with the latter being not only a contingent aggregate, but an indispensable system for the complete identity of conditioned knowledge.

APPENDIX A: TIMELINE

1666 **1** *Gottfried Wilhelm Leibniz* described the philosophy of the whole and its parts in the work DISSERTATIO DE ARTE COMBINATORIA.

1764 **2** *Johann Heinrich Lambert* laid the roots of systems thinking in the work NEUES ORGANON.

1931 **3** Kurt Godel published his INCOMPLETENESS THEOREM, considered to be the benchmark for transdisciplinary thinking.

1972 **4** *Jean Piaget* officially coined the term TRANSDISCIPLINARITY during the L'INTERDISCIPLINARITÉ event, focused on disciplinary relations.

1972 **5** *Erich Jantsch* introduced the first concept of TRANSDISCIPLINARITY with a social purpose during the event L'INTERDISCIPLINARITÉ.

1979 **6** *Joseph J. Kockelmans* published his work INTERDISCIPLINARITY AND HIGHER EDUCATION, which highlighted transdisciplinarity in higher education.

1987 **7** Foundation of *The International Center for Transdisciplinary Research* (CIRET).

1989 **8** *Werner Heisenberg* published REALITY AND ITS ORDER, produced between 1941 and 1942.

1991 **9** *Edgar Morin* published ON COMPLEXITY, with the Dialogic, Organizational Recursion and Holographic principles of complexity.

1994 **10** *Michael Gibbons* et al. released THE NEW PRODUCTION OF KNOWLEDGE, which reinforces the participation of non-academic actors in science.

1994 **11** THE CHARTER OF TRANSDISCIPLINARITY was released during the *First World Congress on Transdisciplinarity* in Portugal.

1996 **12** *Elinor Ostrom* used the term co-production in the transdisciplinary context, in her article CROSSING THE GREAT DIVIDE.

1996 **13** *Basarab Nicolescu* published the MANIFESTO OF TRANSDISCIPLINARITY, with the axioms of transdisciplinarity: ontology, logic and complexity.

2000 **14** The INTERNATIONAL TRANSDISCIPLINARITY CONFERENCE was held in Switzerland, with works based on *Mode 2 knowledge production.*

2004 **15** *Sheila Jasanoff* described the co-production of scientific knowledge in the book STATES OF KNOWLEDGE.

2008 **16** *Gertrude Hadorn* et al. associated transdisciplinary research with the principles of the common good, in their HANDBOOK OF TRANSDISCIPLINARY RESEARCH.

2012 **17** *Mathias Bergmann* et al. classified integrative research in METHODS FOR TRANSDISCIPLINARY RESEARCH.

2013 **18** *Julie Thompson Klein* published the article THE TRANSDISCIPLINARY MOMENT(UM) on the five major keyword clusters of transdisciplinarity.

2014 **19** *Robert Frodeman* used the term Transdisciplinarity of co-production in his book SUSTAINABLE KNOWLEDGE: A THEORY OF INTERDISCIPLINARITY.

2015 **20** *Merritt Polk* used the term Transdisciplinary Co-production in her article TRANSDISCIPLINARY CO-PRODUCTION.

XVII	**XVIII**	**XIX**
'60	'60	
1666	1764	
1	2	

XX						
'00	'10	'20	**'30**	'40	'50	'60
			1931			
			3			

XX									
'70									
1970	1971	**1972**	1973	1974	1975	1976	1977	1978	**1979**
		4, 5							**6**

XX									
'80									
1980	1981	1982	1983	1984	1985	1986	**1987**	1988	**1989**
							7		**8**

XX										
'90										
1990	1991	1992	1993	**1994**	1995	**1996**	1997	1998	1999	**2000**
	9			**10** **11**		**12** **13**				**14**

XXI								
'00								
2001	2002	2003	**2004**	2005	2006	2007	**2008**	2009
			15				**16**	

XXI										
'10										
2010	2011	**2012**	**2013**	**2014**	**2015**	2016	2017	2018	2019	2020
		17	**18**	**19**	**20**					

THE AUTHORS

Lillian Maria Araujo de Rezende Alvares

Professor at the School of Information Science, University of Brasilia (UnB). PhD in Information Science (University of Brasilia) and in Sciences de l'Information et de la Communication (Université de Toulon). Deputy leader of the Research Group on Organizational and Competitive Intelligence (University of Brasilia), member of the Research Group on Integration Engineering and Multilevel Governance of Knowledge and Learning (Federal University of Santa Catarina) and of the Open Access to Scientific Information Research Group (Brazilian Institute of Information in Science and Technology). Author of several articles published in Brazilian and International conference proceedings, journals, books and book chapters. Her major research interests are information management, knowledge management and competitive intelligence. Member of the Science and Technology Libraries Section Standing Committee of the International Federation of Library Associations and Institutions (IFLA) with a term of office until 2025. She was at the Brazilian Institute of Information in Science and Technology (Ibict), a research unit of the Ministry of Science, Technology and Innovations (2014 to 2018).

Patricia de Sá Freire

Professor at the Department of Engineering and Knowledge Management, Federal University of Santa Catarina. PhD in Engineering and Knowledge Management (Federal University of Santa Catarina). Leader of the Research Group on Integration Engineering and Multilevel Governance of Knowledge and Learning, member of the Center for Intelligence, Management and Technology for Innovation and of the Learning and Organizational Memory Group, all from of the Federal University of Santa Catarina. Executive and scientific editor of the International Journal of Knowledge and Management (IJKEM). Her major research interests are the new engineering models of integration and multilevel governance of knowledge and organizational learning and their mechanisms of networked corporate university, neolearning, communication and dialogic leadership, organizational learning networks and collaborative management. For more than 30 years she has been a consultant for the management of strategic and people changes. For two years in a row (2011 and 2012), she was acclaimed by the MAKE Award Brazil as one of the five executives of excellence in Knowledge Management in Brazil. She has authored seven books and organized another six, and published over 300 scientific articles.

INDEX

Lightning Source UK Ltd.
Milton Keynes UK
UKHW041539251122
412842UK00004B/12